# Mathematical Foundations of Information Retrieval

# MATHEMATICAL MODELLING:
## Theory and Applications

### VOLUME 12

This series is aimed at publishing work dealing with the definition, development and application of fundamental theory and methodology, computational and algorithmic implementations and comprehensive empirical studies in mathematical modelling. Work on new mathematics inspired by the construction of mathematical models, combining theory and experiment and furthering the understanding of the systems being modelled are particularly welcomed.

Manuscripts to be considered for publication lie within the following, non-exhaustive list of areas: mathematical modelling in engineering, industrial mathematics, control theory, operations research, decision theory, economic modelling, mathematical programmering, mathematical system theory, geophysical sciences, climate modelling, environmental processes, mathematical modelling in psychology, political science, sociology and behavioural sciences, mathematical biology, mathematical ecology, image processing, computer vision, artificial intelligence, fuzzy systems, and approximate reasoning, genetic algorithms, neural networks, expert systems, pattern recognition, clustering, chaos and fractals.

Original monographs, comprehensive surveys as well as edited collections will be considered for publication.

# Mathematical Foundations of Information Retrieval

*by*

**Sándor Dominich**
*University of Veszprém,*
*Department of Computer Science,*
*Veszprém, Hungary*
*and*
*Buckinghamshire Chilterns University College,*
*High Wycombe, United Kingdom*

**SPRINGER-SCIENCE+BUSINESS MEDIA, B.V.**

A C.I.P. Catalogue record for this book is available from the Library of Congress.

ISBN 978-94-010-3819-5     ISBN 978-94-010-0752-80 (eBook)
DOI 10.1007/978-94-010-0752-8

*Printed on acid-free paper*

To my parents Jolán and Sándor

To my wife Emőke, and our daughter Emőke

# Contents

# Preface

## Background

Information Retrieval (*IR*) has become, mainly as a result of the huge impact of the World Wide Web (WWW) and CD–ROM industry, one of the most important theoretical and practical research topics in Information and Computer Science. Since the inception of its first theoretical roots about 40 years ago, *IR* has made a variety of practical, experimental and technological advances. It is usually defined as being concerned with the organisation, storage, retrieval and evaluation of information (stored in computer databases) that is likely to be relevant to users' information needs (expressed in queries). A huge number of articles published in specialised journals and at conferences (such as, for example, the *Journal of the American Society for Information Science, Information Processing and Management, The Computer Journal, Information Retrieval, Journal of Documentation, ACM TOIS, ACM SIGIR Conferences*, etc.) deal with many different aspects of *IR*. A number of books have also been written about *IR*, for example: van Rijsbergen, 1979; Salton and McGill, 1983; Korfhage, 1997; Kowalski, 1997; Baeza–Yates and Ribeiro–Neto, 1999; etc.. *IR* is typically divided and presented in a structure (models, data structures, algorithms, indexing, evaluation, human–computer interaction, digital libraries, WWW–related aspects, and so on) that reflects its interdisciplinary nature.

All theoretical and practical research in *IR* is ultimately based on a few basic models (or types) which have been elaborated over time. Every model has a formal (mathematical, algorithmic, logical) description of some sort, and these decriptions are scattered all over the literature. A book, containing a consistent and comprehensive formal mathematical description of the different methods of retrievals along with their underlying abstract mathematical structures in the basic *IR* models, as well as a unified mathematical foundation of *IR* allowing for a formally built unified treatment of all the mathematical results achieved so far in *IR*, is lacking. This lack is particularly recognised, on the one hand, by a wide range of *IR* researchers who would like to have an overview and comprehensive formal description of the different *IR* models in one volume, and on the other hand, by newcomers to the field — even including a potential scientific audience which has been excluded so far, namely, coming from formal disciplines such as mathematical, electrical and related disciplines, and by those who would like to teach *IR* to a more diverse scientific audience oriented towards formal methods, too, or to do formal theoretical research in *IR*.

**Goal of the Book**

The present book primarily aims at helping fill this gap — it gives formal mathematical descriptions of the retrievals in the basic *IR* models, in a unified mathematical style, format, and language, and it creates a consistent mathematical framework within which major mathematical results, achieved so far in *IR*, are included and treated. Thus this book creates a unified introduction to the mathematical foundations of *IR*. The book can also help *IR* researchers by offering a broad ranging view of the different *IR* models and results in one volume. Thus *IR* receives a mathematical individuality, and becomes a mathematical discipline, too (beside — as is well known and widely accepted — belonging to the social, computer, and informational sciences as well).

**Organisation of the Book**

In order to help the reader avoid looking for different mathematics books, Chapter 2 contains, in handbook form, the core mathematical knowledge which *IR* models rely on. Definitions, properties (without proofs) and examples are given. This chapter can, obviously, be skipped when an appropriate mathematical background has already been built, or it can be consulted just to refresh some of the mathematical knowledge. At the same time this chapter can also be used as a companion in case this mathematical knowledge, or parts of it, is to be taught (for example, as a complementary subject for *IR* studies).

Chapter 3 contains the decriptions of the basic *IR* models. A representative model is described in each case in a consistent mathematical fashion, along with a simple example to ease understanding. This chapter can be read and taught independently, too (Chapter 2 can be consulted if need be).

Chapter 4 contains a mathematical foundation and unified mathematical theory of *IR* as a mathematical discipline of *IR*. All mathematical results achieved so far in *IR* are included. This chapter, too, can be read and taught independently; it does not necessarily require Chapter 3, but it is assumed that some of the knowedge of Chapter 2 is already at hand.

Chapter 5 is concerned with elaborating a mathematical foundation and theory for the relevance effectiveness in *IR*. It, too, can be read and taught separately.

Finally, Chapter 6 deals with a few special topics, such as *IR* and Decision Theory, and Fusion. These can be used as a starting point for further research.

The book ends with Appendices containing complex examples and procedures for the vector, probabilistic, and interaction *IR* models.

## Intended Audience

The book is intended as a reference book for *IR* professionals (researchers, lecturers, PhD students, programmers, systems designers, and independent study), but also for newcomers, even from strongly formally oriented fields such as, e.g., mathematics.

The book can also be used as a textbook for a one or two semester course in *IR*, for example in an undergraduate course, a graduate or advanced graduate course, designed, for instance, for IT, Computer Science, Mathematics, Electrical Engineering students.

It would be helpful — although not necessary — for those interested to have a good affinity with formal methods and an algorithmic view as well.

An understanding of a high level programming language or of some Computer Algebra software, such as, e.g., MathCAD, would be advantageous for understanding of complex examples and procedures given in the Appendices.

## Methodology of the Book

The methodology in writing this book is as follows.

All basic mathematical knowledge necessary for and used in *IR* has been carefully collected, and is systematically presented in Chapter 2, which thus helps view what is needed from Mathematics and how deep. All concepts, properties, and theorems are methodically presented along with examples for ease of understanding. Proofs are omitted in this chapter, since these can be found in mathematics books, and would go beyond the scope of this book. The chapter ends with a carefully selected mathematical bibliography to help the interested reader.

The basic *IR* model types have been carefully collected, and are methodically described in Chapter 3 according to the following scheme: a formal mathematical description of a representative model is given along with a simple example. Each model can also be read and taught independently.

The mathematical foundations and mathematical theory of classical *IR* can be found in Chapter 4, which adopts an axiomatic style. All results can be nicely derived formally from one common definition. The whole chapter can be read and taught independently, too — as a mathematical discipline of *IR*. It is not a requirement to understand the particular models first — as these are formally derived within this theory. Even within the chapter, some parts can be read and taught separately as well: binary *IR*, non–binary *IR*, probabilistic *IR*, interaction *IR*. This chapter, too, can be read and taught independently.

The *IR* references appear at the end of the book. Each part treating a topic ends with bibliographic remarks.

There are three appendices to this book. The first appendix contains a complex example along with procedures for the vector space model of *IR*. Appendix 2 contains a complex example along with procedures for the probabilistic model of *IR*, whilst the third appendix presents a complex example for the interaction model of *IR*.

Many parts of this book, as, for example, indicated above, are so written that they can be read and taught separately — this should help to use the book in a variety of ways and by different categories of readers. However, the book is not just a simple collection of standalone parts: a complete 'linear' (i.e., from the beginning to the end) reading of the book is, of course, also possible, and this is highly recommended in order to aquire a complete, articulate, and colourful picture of all formal and mathematical aspects of *IR*, and also in order to place *IR* within the context of exact sciences.

## Paths through the Book

This book can be used in a number of ways in several areas including, for example, Computer Science, Formal Methods, Information Science, Mathematics.

There are already standard courses in *IR* in many teaching programmes around the world. This book can be used to complement these courses, and the following list contains a few suggestions:

— **Information Retrieval** (Computer Science, undergraduate): this course can be complemented with most of Chapter 3.

— **Advanced Information Retrieval** (Computer Science, graduate): this course can be complemented with Chapters 3 and 5, and the complex examples and procedures in the Appendices.
— **Information Retrieval** (Information Science, undergraduate): this course can be complemented with the examples of Chapter 3.
— **Information Retrieval** (Library Science, undergraduate, graduate): this course can be complemented with the Introduction and relevant examples of Chapter 3.
— **Topics in *IR*** (Computer Science, undergraduate): this course can be complemented with, for example, the procedures in the Appendices.
— **Topics in *IR*** (Computer Science, graduate): this course can be complemented with Chapter 6.
— **Topics in Web *IR*** (Computer Science, undergraduate, graduate): this course can be complemented with e.g. the fusion topic (Chapter 6).

This book can also be used to offer new and/or special courses to other scientific audiences as well. The following list contains a few suggestions:

— **Information Retrieval** (Technical Informatics, undergraduate): such a course should include Chapter 3 and Appendices.
— **Advanced Information Retrieval** (Technical Informatics, graduate): such a course should include Chapters 4, 5 and 6 and Appendices.
— **Information Retrieval** (Mathematics, undergraduate): such a course should include Chapter 4.
— **Advanced Information Retrieval** (Mathematics, graduate): such a course should include Chapters 4 and 6.

Further, the book can be used as a starting point in or to complement different (e.g., M.Sc., Ph.D.) research programmes. The following list contains a few suggestions:

— **Mathematical Models and Structures in *IR***: starting topics should include Chapters 3 and 5, and Appendices.
— **Mathematical Theory of *IR***: Chapters 4 and 5.
— **Models in *IR*** : starting topics should include Chapter 3.
— **Artificial Intelligence Applications**: starting topics should include relevant parts of Chapter 3.
— **Applied Functional Analysis**: starting topics should include relevant parts of Chapter 4 and Appendices.
— **Applied Abstract Algebra**: starting topics should include relevant part of Chapter 4.

— **Applied Recursion Theory**: starting topics should include Chapter 5, along with relevant parts of Chapter 4.

— **Applied Matroid Theory**: starting topics should include relevant parts of Chapter 4.

In general, this book is also a helpful tool for those whishing to have a formal and unified overview of current *IR* models and theoretical research, as well as for those wishing to aquire the mathematical foundations of *IR* as a formal mathematical discipline.

## Corrections, Suggestions

As a book of this nature is vulnerable to errors, disagreements, omissions, your input is kindly solicited and opinions are most welcome for possible further reprints and editions, to the following address: Kluwer Academic Publishers, or e–mail: *sdomin01@buckscol.ac.uk* or *dominich@dcs.vein.hu*.

Sándor Dominich
*Fertőhomok*

# ACKNOWLEDGEMENTS

Special thanks must go first to my family, the two Emőke, who over the years have been very understanding and supportive during efforts and concentrations of writing this book.

Many thanks go to the Faculty of Technology, Buckinghamshire Chilterns University College, High Wycombe, United Kingdom, and to my Colleagues in the Department of Information Technology and Computer Science, for their support and fruitful discussions we had at the time I began to formulate and write down the mathematical foundations of $IR$. I especially thank Dorette Biggs, Frederick Corbett, Mark Day, Martin Hamer, Milton Munroe, Trevor Nicholls, and Janet Payne.

Next, I would like to thank the Department of Computer Science, University of Veszprém, Hungary, and especially Ferenc Friedler, for the highly supportive environment provided during periods of writing this book.

Special thanks go to final year undergraduates in Computer Science at the Buckinghamshire Chilterns University College, High Wycombe, United Kingdom, as well as to PhD students in Computer Science of University of Veszprém, Veszprém, Hungary, for being the audience of a course where we tested — with a positive outcome — that the mathematical theory of $IR$, as in Chapter 4, can be taught independently, too, as a formal discipline, and without any previous knowledge of $IR$.

Very special thanks go to Rudolf Keil, Don Kraft, Mounia Lalmas, Ádám Nagy, Gyula Maurer, Gabriella Pasi, Gábor Prószéky, Keith van Rijsbergen, Lajos Rónyai, Tamás Roska, Alan Smeaton, Mária Horváthné Szilárd, Péter Szolgay, and many others, for their helpful comments and suggestions in the period of writing this book.

I also thank Ádám Berkes, Gábor Dröszler, János Forman, Zoltán Jeges and Zoltán Papp who helped to write and test the procedures.

Many thanks go to the National Science Foundation (OTKA), Hungary, and Pro Renovanda Cultura Hungariae Foundation, Hungary, for their financial support of related research.

I am grateful to the anonymous reviewers for their useful comments and generous help.

Last but not least, I am indebted to Kluwer Academic Publishers for making the publication of this book possible.

Sándor Dominich
*Fertőhomok*

# Chapter 1

# INTRODUCTION

## 1. INFORMATION RETRIEVAL

*Information Retrieval* (*IR*) is concerned with the organisation, storage, retrieval, and evaluation of information relevant to a user's information need. Figure 1 is a schematic illustrating the concept of *IR*. The main components of *IR* are as follows:

1. user;
2. information need;
3. request;
4. query;
5. information stored in computer(s);
6. appropriate computer programs.

The user (e.g., researcher, librarian, businessman, tourist, etc.) has an information need (i.e., wants to find out something, is looking for information on something; e.g. articles published on a certain subject, books written by an author, banks offering online banking services, travel agencies with last minute offers, etc.). The information need is formulated in a request for information, in natural language. The request is then expressed in the form of a query, in a form that is required by the computer programs (e.g., according to the syntax of a query language). These programs retrieve information in response to a query, e.g., they return database records, journal articles, WWW (World Wide Web) pages, etc..

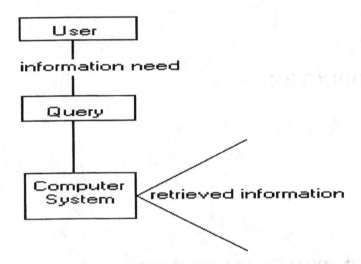

*Figure 1.* Schematic illustrating the concept of Information Retrieval (*IR*): A user has an
information need which is expressed in the form of a query required by a system of computer
programs which retrieve information stored on disks.

This is the reason why, mainly in practice, *IR* can also be viewed as a
system, and the term Information Retrieval System (*IRS*) is also used.

Let us consider an example. CARL (California Association of Research
Libraries) host a large store of objects called UnCover (42 databases and 420
library catalogues containing records describing most journals and their
contents, 4000 citations added daily; telnet access: database.carl.org). If a
user, say *U*, is interested in journal articles and/or authors on, e.g.,
'mathematical methods and techniques used in information retrieval' then
this is the user's information need; let us denote it by *IN*, thus:

*IN* = mathematical methods and techniques used in information retrieval.

This information need *IN* is re-formulated in a form accepted by the
processor of CARL; it thus becomes a query, say *Q*, e.g., as follows:

*Q* = INFORMATION RETRIEVAL MATHEMATICS.

The CARL processor looks for objects (articles, papers) containing the word
RETRIEVAL and finds 3318 objects. It then searches these 3318 objects to
find those containing both RETRIEVAL and MATHEMATICS. Finally, it

finds two objects containing all three terms: RETRIEVAL and MATHEMATICS and INFORMATION. The titles, authors, dates, names, journal titles for these two objects are displayed as CARL's answer to query $Q$.

Information is stored in computer databases. More generally, information is stored in entities which may be generically referred to as objects $O$, e.g., abstracts, articles, images, sounds, etc.; these are traditionally called documents. The objects should be suitably represented, in such a way that they can be subjected to appropriate algorithms and computer programs. The same holds for queries, too. Thus our schematic becomes as shown in Figure 2.

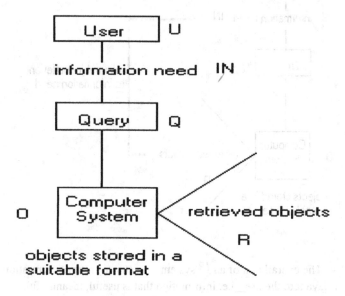

*Figure 2.* Enhanced schematic of *IR* with notations. Information is retrieved from information stored in computers (on disks) in a suitable format of objects (traditionally called documents).

The overall aim of an *IR* system is to or try to return information which is relevant to the user, i.e. information that is useful, meaningful.

If the two answers returned by the CARL processor are in English and the user speaks English, then the retrieved objects (which may be generically denoted by, say, $R$), are — or more exactly may be — useful answers. If, however, the user does not speak English, then these answers are meaningless to him/her. In other words, the retrieved objects may or may not be relevant to the user's information need. Thus, the schematic can be further enhanced as shown in Figure 3.

Thus *IR* may be re–formulated symbolically — or formally — as a 4–tuple yielding retrieved objects as follows:

$$IR = (U, IN, Q, O) \rightarrow R$$

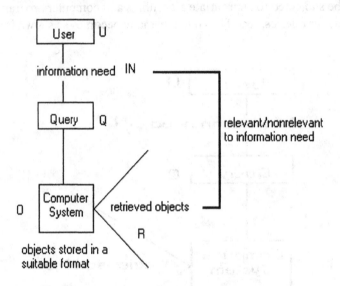

*Figure 3.* The overall aim of an *IR* system is to (try to) return information that is relevant to the user, i.e. information that is useful, meaningful.

The information need *IN* is more than its expression as a query *Q*: *IN* comprises query *Q* plus additional information about user *U*. This additional information is specific to the user: spoken languages, fields of interest, preferred journals, specialisation, profession, most frequently used queries, etc.. The importance of additional information consists in that it is one factor in judgement of relevance, when judging whether a retrieved object is relevant or not. For example, the same search term PROGRAM has different meanings for a computer programmer (meaning a text written in the C programming language and solving a differential equation) and for a conference organiser (meaning a structure and sequence of scientific and social events during the conference). The additional information is obvious for the user (he/she implicitly assumes it) but not for the computer. Thus we may term this additional information as being an implicit information $I$ specific to the user $U$, and we may write:

$$IN = (Q, I)$$

Thus the meaning of the concept of $IR$ can be re–formulated as being concerned with finding an appropriate — relevance — relationship, say $\Re$, between objects $O$ and information need $IN$; symbolically:

$$IR = \Re(O, IN) = \Re(O, (Q, I))$$

In order for an $IR$ system to find such a relation $\Re$ it should be made possible to take into account the implicit information $I$ as well, and ideally the information which can be deduced (inferred) from $I$ to obtain as complete a picture of user $U$ as possible. Thus finding an appropriate relation $\Re$ would mean obtaining (deriving, inferring) those objects $O$ which match the meaning of the query $Q$ and satisfy the implicit information $I$. With these $IR$ becomes:

$$IR = \Re(O, (Q, \langle I, \longmapsto \rangle))$$

where $\langle I, \longmapsto \rangle$ means $I$ plus information derivable (e.g. in some language or logic) or inferred from $I$. Of course, the relation $\Re$ is established with some (un)certainty $m$; thus:

$$IR = m[\Re(O, (Q, \langle I, \longmapsto \rangle))]$$

## 2. INFORMATION RETRIEVAL MODELS

The last decade, and especially the last few years, mainly as a result of the impact of the Internet, WWW, and CD–ROM industry, has been characterised by a boost in research of theoretical modelling in Information Retrieval (beside intensive practical research).

Several different $IR$ models have been elaborated so far. From a purely formal mathematical point of view they differ from each other in the way objects (documents and queries) are represented and in which retrieval itself is conceived (modelled, defined). For example, in the vector space model of $IR$, objects (texts) are represented as strings of numbers (called vectors), denoting importances of keywords in texts. Retrieval is defined using numerical relationships (a distance meant to express a 'closeness') between strings. In another model of *IR*, called Information Logic *IR*, retrieval is viewed as an inference process from documents to queries.

These models together with other models of *IR* are described in Chapter 3.

## 3.   MATHEMATICS IN INFORMATION RETRIEVAL MODELS

The first models of *IR*, and virtually all commercial *IR* systems today, are based on the classical models of *IR*. These can be symbolically written as follows:

$$IR = \Re(O, Q)$$

In other words, both object (document) $O$ and query $Q$ are formally represented as mathematical objects of the same type, such as, for example, vectors of numbers, so that they can be subjected to different mathematical operations so that the relationship $\Re$ takes a computable form (a distance between vectors).

Chapter 4 contains a unified mathematical foundation and theory of the *IR* models.

The Boolean model of *IR* is used in virtually all commercial *IR* sytems today. In this model documents are conceived as sets of keywords. The query consists of keywords connected by logical operators (e.g. logical AND, OR). In order to answer the query, those documents are located first which contain a query keyword, and then the corresponding set operations (for example, set union corresponds to logical OR) are applied on the documents thus located in order to find a final answer to the query. In spite of its simplicity, the Boolean model of *IR* bears nice and surprising mathematical properties — these are dealt with at the end of Chapter 4.

Another basic model of *IR* is the Vector Space Model. This is based on the mathematical concept of a vector. For example, let us assume that we have four objects (documents) as follows:

1. $o_1$ = *Bayes' Principle: The principle that in estimating a parameter one should initially assume that each possible value has equal probability (a uniform prior distribution).*

2. $o_2$ = *Bayesian Conditionalisation: This is a mathematical procedure with which we revise a probability function after receiving new evidence. Let us say that we have probability function P(.) and that through observation we come to learn that E is true. If we obey this rule our new probability function, Q(.), should be*

*such that for all X Q(X)=P(X|E); we are then said to have "conditionalised on E".*

3. $o_3$ = *Bayesian Decision Theory: A mathematical theory of decision making which presumes utility and probability functions, and according to which the act to be chosen is the Bayes act, i.e. the one with highest Subjective Expected Utility. If one had unlimited time and calculating power with which to make every decision, this procedure would be the best way to make any decision.*

4. $o_4$ = *Bayesian Epistemology: A philosophical theory which holds that the epistemic status of a proposition (i.e., how well established it is) is best measured by a probability, and that the proper way to revise this probability is given by Bayesian conditionalisation or similar procedures. A Bayesian epistemologist would use probability to define, and explore, the relationship between concepts such as epistemic status, support, or explanatory power.*

Each object $o_i$ is assigned a vector $\mathbf{v}_i$ as follows:

$$\mathbf{v}_1, \mathbf{v}_2, \mathbf{v}_3, \mathbf{v}_4$$

For example, a vector $\mathbf{v}_i$ associated to object $o_i$ might be the following vector (string of numbers)

$$\mathbf{v}_i = (v_{i1}, v_{i2}, ..., v_{ij}, ..., v_{iN})$$

where $N$ denotes the number of predefined index terms (keywords)

$$t_1, t_2, ..., t_j, ..., t_N$$

used to identify the objects, and $v_{ij}$ is the number of occurrences of the index term $t_j$ in object $o_i$. For example, let the index terms be as follows ($N = 3$):

1. $t_1$ = Bayesian Conditionalisation
2. $t_2$ = probability function
3. $t_3$ = probability

The associated vectors (strings of numbers) are as follows:

$$
\begin{array}{l}
\quad\; t_1\; t_2\; t_3 \\
\mathbf{v}_1 = (0, 0, 1) \\
\mathbf{v}_2 = (1, 3, 3) \\
\mathbf{v}_3 = (0, 1, 1) \\
\mathbf{v}_4 = (1, 0, 3)
\end{array}
$$

The vectors

$$\mathbf{v}_1, \mathbf{v}_2, \mathbf{v}_3, \mathbf{v}_4$$

can be represented as points

$$P_1, P_2, P_3, P_4$$

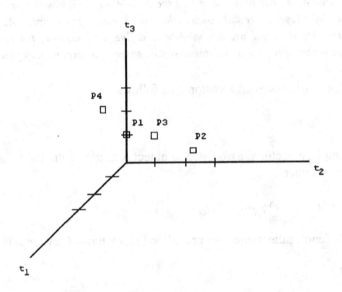

*Figure 4.* Objects are represented as points (vectors, strings of numbers) in a Euclidean space, in the vector model of *IR*.

in a (multi–dimensional) Euclidean space where the coordinates of point $P_i$ are the components of $\mathbf{v}_i$, i.e. $P_i(v_{i1}, v_{i2}, ..., v_{ij}, ..., v_{iN})$. Taking our example we obtain the points shown in Figure 4.

Given a query $Q$, it is represented the same way as any other object. Thus let the query be, say, $Q = P_2$. Those objects are considered to be answers (retrieved in response) to $Q$ which are 'close' enough to it. This 'closeness' should be modelled mathematically. One idea is that the more index terms two objects have in common the more similar in meaning they are/should be. Thus the likeness, or similarity, of objects is inversely proportional to, e.g., the geometric (Euclidean) distance between them: the more they have in common (in meaning, expressed by index terms) the closer they are to each other (geometrically). The distance between $Q$ and $P_1$ is $((0-1)^2+(0-3)^2+(1-3)^2)^{1/2} = \sqrt{14}$, between $Q$ and $P_3$ is $((0-1)^2+(1-3)^2+(1-3)^2)^{1/2} = \sqrt{9}$, between $Q$ and $P_4$ is $((1-1)^2+(0-3)^2+(3-3)^2)^{1/2} = 3$. Thus, in our example, $P_3$ and $P_4$ are closest to $Q$, and hence retrieved in response to $Q$.

Obviously, it seems reasonable to retrieve not just the closest object, but all objects which are within a pre–defined distance, say, $d$, usually called a threshold, i.e., whose distance from $Q$ is less than $d$. For example, if $d = 3$, then two objects are retrieved: $P_3$ and $P_4$.

Thus from a purely mathematical (formal) point of view, objects are situated at certain distances from each other. Those objects are retrieved in response to query $Q$ which are closest or close enough (within a threshold) to $Q$, where 'closest' or 'close enough' is expressed in terms of a threshold: those objects are retrieved whose distance from $Q$ does not exceed a pre–defined threshold, say, $d$. In other words, retrieval means defining a neighbourhood (vicinity) $V$ of query $Q$ in which all objects (points) are closer to $Q$ than a pre–defined threshold $d$ (Figure 5).

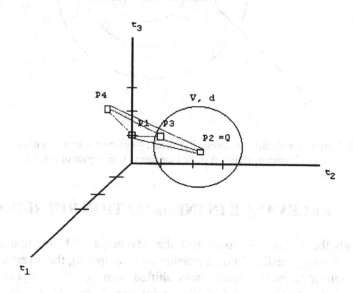

*Figure 5.* Retrieval, in the vector mdoel of *IR,* means defining a $d$–diameter neighbourhood $V$ of query: objects situated within a 'distance' $d$ around query $Q$.

By taking different values for $d$, e.g., $d_1$, $d_2$, ..., different neighbourhoods $V$, e.g. $V_1$, $V_2$, ..., of the same query $Q$ are obtained (Figure 6). Thus the process of retrieval may be conceived as defining a series of neighbourhoods $V$ (vicinities) of query $Q$. From a purely mathematical point of view, therefore, the underlying abstract mathematical structure of the vector space model of *IR* is defining a special series of vicinities (mathematically, a topology). The mathematical theory of the classical *IR* models, in general, is contained in Chapter 4.

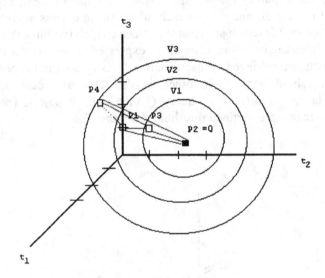

*Figure 6.* Retrievals with different threshold values $d$ — in the vector model of *IR* — define a special series of vicinities (called topology) in the space of objects.

## 4.      RELEVANCE IN INFORMATION RETRIEVAL

Although the vector *IR* model had the advantage of being immediately computationally feasible (programmable on a computer), the overall aim of *IR* (to return relevant objects) was shifted somehow to peripherals. To overcome this another model of *IR* was developed, the probabilistic model. The idea was to estimate a degree of relevance of an arbitrary object to the query based on a feedback from the user. Formally, the degree of relevance is another form of 'distance'.

In the probabilistic model retrieval takes place in two phases, in fact: a) in the first phase an initial set of objects is retrieved, e.g., using the vector space model; b) relevance feedback (subjective momentum): based on the user's indication, distances are re–calculated and retrieval is repeated.

a) In the first phase, objects, say, $P_3$ and $P_1$, are retrieved using, e.g., the vector model (Figure 7).

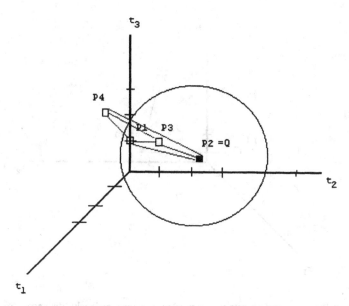

*Figure 7.* In the probabilistic model, where typically there is relevance feedback, an initial set of objects is retrieved first, e.g., using a vector model.

b) Relevance feedback (subjective momentum). The user indicates which of the retrieved objects are relevant and which are irrelevant. Let us assume that $P_1$ is relevant and $P_3$ is irrelevant. Based on this feedback from the user, 'distances' from $Q$ to the other objects are re–calculated (they express a degree of relevance/irrelevance to $Q$). Thus the initial objects structure is (partially) re–shaped (Figure 8).

The objects are ranked in decreasing order of their new distances (relevance), say $P_1$, $P_4$, $P_3$..

A cut off value can be used in order to avoid retrieving the whole store of objects. Thus, say, $P_1$ and $P_4$ can be retrieved.

The process of repeatedly applying relevance feedback can be performed, in principle, as many times as desired, which yields a sequence of sets of retrieved objects:

$R_1, R_2, R_3, ...$

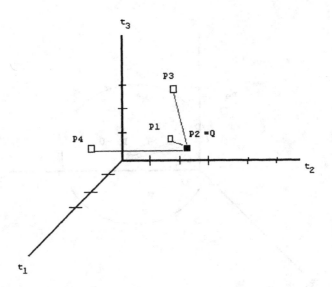

*Figure 8.* Re–structured objects structure after relevance feedback: 'distances' (representing probabilities of relevance/nonrelevence) from query to the other objects are changed.

where, for instance,

$R_1 = \{P_1, P_3\}$ is an initial set of retrieved objects,

$R_2 = \{P_1, P_4\}$ is a set of retrieved objects after the first relevance feedback,

and so on. This sequence, albeit it possibly — in principle — be infinite, is not an arbitrary structure nor is it chaotically generated: every retrieved set $R_i$ of objects is used in the next relevance feedback yielding the next $R_{i+1}$ (Figure 9).

Thus from a purely formal mathematical point of view, because of this recursive property, the sequence $(R_i)$ of sets of retrieved objects form a special mathematical structure (called recursively enumerable set) which thus characterises relevance feedback. Its mathematical theory is described in Chapter 5.

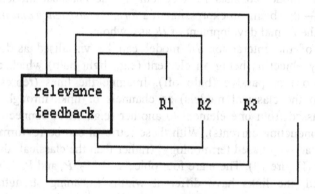

*Figure 9*. Recursive nature of relevance feedback in the probabilistic model: every retrieved set $R_i$ of objects is used in the next relevance feedback, which yields a next set of retrieved objects $R_{i+1}$.

## 5. INTERACTION INFORMATION RETRIEVAL

In both the vector and probabilistic types of model the question of whether or not to retrieve an object is based, in point of fact, on a 'distance' (e.g., Euclidean, degree of relevance).

In the vector model answering a query does not modify the objects structure, i.e., the distances between them do not change. In the probabilistic model some of the distances change after relevance feedback and before a new retrieval. The idea of changeable distances is further developed in the

Interaction model of *IR* (based on the Copenhagen Interpretation in Quantum Logic):

> *The objects are effectively interconnected with each other. The query, before being answered, becomes a member of this structure of interconnected objects as if it were just another object. Those objects are said to be retrieved in response to the query which are memories recalled by an activation spreading started at the query.*

From a purely formal mathematical point of view, one of the niceties of the Interaction *IR* model consists in it containing the classical models as a special case — this being an expression of a *formal consistency and internal dynamics* in the formal development of *IR* as a whole.

The idea of the Interaction *IR* model can be visualised as follows. Conceive any object as being an element (e.g., light bulb) which can be active (bulb on) or passive (bulb off). Imagine the lines (representing 'distances' in the classical models) as channels or links through which activity is passed from one element to another (e.g., wires connecting the bulbs, and conducting currents). With these retrieval can be re-formulated using a new 'activity based terminology' (rather than the classical 'distance terminology' (Figure 10). There are four objects: $P_1$, $P_2$, $P_3$ and $P_4$. They are interconnected, the links have different widths (meaning strengths or weitghts). First (Figure 10a), $P_2 = Q$ is active (on; white object) and all the others are inactive (off; dark objects). A spreading of activation is started at $P_2$ and it spreads along the widest connection (largest strength) towards $P_4$ which thus becomes active, whilst all the other objects become inactive (Figure 10b). This process of activation spreading continues (Figure 10b—e) until a circle forms (Figure 10f), i.e., a situation where the activation goes in circle through the same objects; in our example $P_1$ and $P_3$. This circle acts as a (short term) memory recalled by the activation spreading started at $P2$. Thus, $P_1$ and $P_3$ were retrieved.

Retrieval interpreted as recalled memories is a keypoint in characterising, from a purely mathematical point of view, the interaction model. The activation is started at $Q$ and spread over through the network until recalled memories are formed in which the activation goes in circle (Figure 11). From a formal mathematical point of view the circles (together with the other connections) means defining an interesting and specific mathematical structure (called a matroid) within the network of elements.

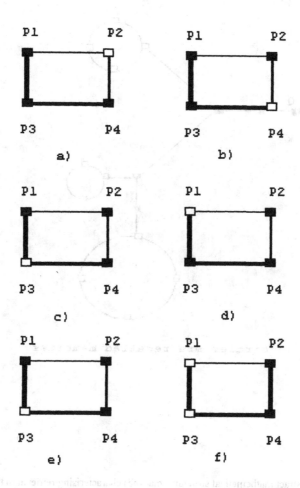

*Figure 10.* Interaction *IR.* Objects are interconnceted. White objects are active, dark ones are inactive. An activation is started at $Q$ (see a)) and spread through the widest connection (see b)—e)) until $P_1$ and $P_3$ activate each other (see f)).

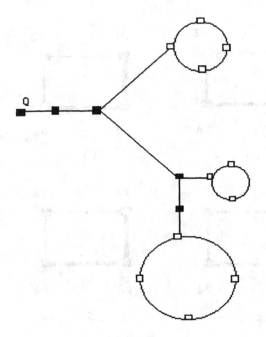

`circles are recalled memories`

*Figure 11.* Abstract mathematical structure (matroid) characterising retrieval in the interaction *IR*. The activation is started at *Q* and spread over through the network until three recalled memories (the circles) are formed. The circles (together with the other connections) means defining an interesting and specific mathematical structure within the network of elements which is called a matroid.

# 6.    MATHEMATICAL FOUNDATION OF INFORMATION RETRIEVAL

The mathematical theory of *IR*, presented in Chapter 4, is based on a unified mathematical definition (the axiomatic style). In what follows, this unified common basis is described using examples.

Let the set of terms be

$T$ = {computer, information, library}

and the set of objects be

$O$ = {JASIS, IPM}.

where JASIS and IPM are the names of two journals: JASIS = *Journal of the American Society for Information Science*, and IPM = *Information Processing and Management*. Each object is represented as a set of pairs

(term, degree_to_which_term_belongs_to_object)

for example, as follows:

JASIS = {(computer, 1), (information, 1), (library, 0.5)}
IPM = {((computer, 0.8), (information, 1), (library, 0.5)}

Or, alternatively, the objects can be represented in the form of a table as follows (Table 1):

*Table 1*. Objects as sets of pairs (fuzzy sets)

|       | computer | information | library |
|-------|----------|-------------|---------|
| JASIS | 1        | 1           | 0.5     |
| IPM   | 0.8      | 1           | 0.5     |

This way of representing the objects (sets of pairs) is achieved by using the concept of fuzzy sets.

Let us assume that there are different criteria, e.g., just one criterion called, say, relevance; or two criteria called, say, relevance and irrelevance; or three criteria called, say, relevance, irrelevance, and undefined. A criterion may be conceived as being a relationship between every pair of objects, i.e., a viewpoint according to which each object is compared to every other object. For example, as follows:

relevance   =   {((JASIS, JASIS), 1), ((JASIS, IPM), 0.5), ((IPM, JASIS),
                        0.5), ((IPM, IPM), 1)}

irrelevance =   {((JASIS, JASIS), 0), ((JASIS, IPM), 0.5), ((IPM, JASIS),
                        0.5), ((IPM, IPM), 0)}

undefined   =   {((JASIS, JASIS), 1), ((JASIS, IPM), 0.3), ((IPM, JASIS),
                        0.2), ((IPM, IPM), 1)}

Thus, mathematically, this is equivalent to conceiving each criterion as being a fuzzy binary relationship. Visually, taking our example, each criterion can be represented in a three–dimensional plot where the X and Y axes show the objects and the vertical axis measures the degree of that criterion. (Figure 12).

A query $Q$ is, formally, just one of the objects. Thus considering $Q$ is equivalent to cutting the corresponding surface with a vertical plane corresponding to $Q$. The result will be a curve in space (the result of cutting the surface with a plane). There are as many curves as criteria, one curve in each criterion plot (Figure 13). In our example the object JASIS is taken as a query, and the corresponding curves are the shown as thick lines. If all curves are represented in one system of coordinates one gets a better visual impression of the general formal definition of classical *IR* (Figure 14). For a better visual impression longer curves are shown. Retrieval is defined relative to one fixed criterion, say $C$, as follows: those objects are retrieved which are above the other query curves and also above a threshold (which can be represented by a horizontal straight line).

Using this general way of conceiving objects, query and criteria, the two classical *IR* models can be obtained as particular cases of it.

The vector model is obtained as follows. The number of criteria is just one; let this criterion be called relevance. This means that there is just one system of coordinates and one query curve. A horizontal straight line represents the threshold value. Those objects are retrieved which are above the straight line (Figure 15).

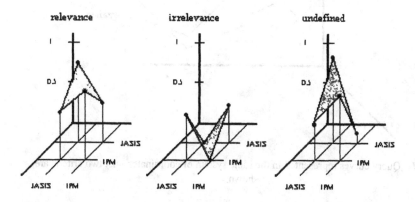

*Figure 12.* Criteria as surfaces (fuzzy binary relationships). See example in text.

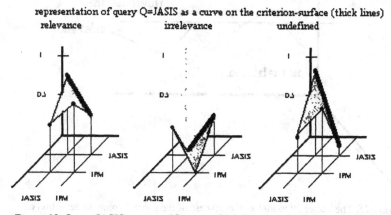

*Figure 13.* Query JASIS as curve (shown as thick lines) on the criterion surface.

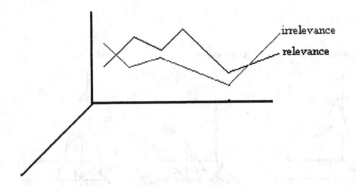

*Figure 14.* Query curves represented in the same system of coordinates. Just two criteria are shown.

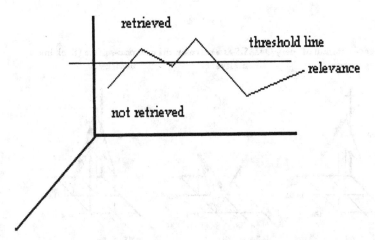

*Figure 15.* The vector *IR* model as special case of the general common definition of classical *IR* (see text)

The probabilistic model of *IR* is obtained as follows. The number of criteria is two; let these two criteria be called relevance and non–relevance. This means two systems of coordinates and two query curves. A horizontal straight line represents the cut–off value. Retrieval is defined with respect to one fixed criterion, relevance. Those objects are retrieved which are above the straight line and the other curve (representing non–relevance; Figure 16).

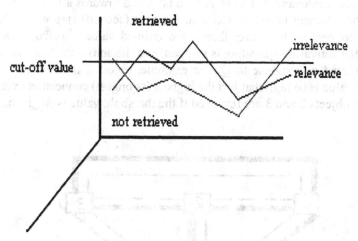

*Figure 16.* The probabilistic *IR* model as a special case of the general common definition of classical IR. (See text)

Thus there is a common general mathematical framework, or superstructure (definition), of which the two classical models, vector and probabilistic, are special (particular) cases.

But is there any connection, in mathematical terms, between the classical models and the interaction model?

It is shown that there is a connection: the interaction model is more general than (is a generalisation of) the classical models. In other words, the general common superstructure of the classical *IR* is a special case of the interaction model.

In the interaction model there are multiple bidirectional connections between objects. Formally, the classical model is a special case of the interaction model as follows. Let the number of connections between any two objects be the same (constant). This number is equal to the number of criteria in classical *IR*, i.e., one (relevance) or two (relevance, non–relevance). Thus between any two objects there is one connection for every criterion. Further, let us fix one of the connections, i.e., one of the criteria (relevance). When an activation is started at object query, it is spread according to the following special rule: that object becomes active whose fixed (see above) input connection exceeds a threshold value. The spreading of activation is stopped after the first step (i.e., the formation of circles is not allowed). This rule is a particular case of the general way in which activation spreads in the interaction model (Figure 17). An activation is started at $Q$. Let the thicker connection be the one which is fixed (it corresponds to relevance). The activation is spread towards all elements 1, 2 and 3. That element becomes active in the next (second) step whose fixed (thicker) connection is thicker than a pre–defined value (threshold). After this step the activation spreading is stopped, and the active elements are said to be retrieved in response to $Q$. For example, object 2 is retrieved if the threshold value is so high that just the highest (strongest) connection exceeds it; or both objects 2 and 3 are retrieved if the threshold value is so given.

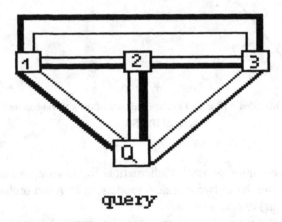

query

*Figure 17*. Classical *IR* as a special cse of interaction *IR*. Activation starts at $Q$, and it spreads towards all objects 1, 2 and 3 through the thicker connections. See text for more details.

# Chapter 2

# MATHEMATICS HANDBOOK

This *MATHEMATICS HANDBOOK* contains the core mathematical knowledge which *IR* relies on.

The basic mathematical concepts and results (theorems, properties, relations) necessary for an understanding of the different models, methods, and techniques used in *IR* are given.

The method adopted in this *HANDBOOK* is as follows: All necessary mathematical concepts and results are presented in a systematic, though concise, manner. Proofs, detailed properties, procedures, methods, multiple definitions, very didactic examples and figures, as well as exercises and problems are omitted, since these would go beyond the role of this book. The interested reader is referred, however, to a carefully selected mathematical bibliography at the end of the *HANDBOOK*, in case the need for more detailed and deeper mathematical knowledge arises.

In some cases, very specific knowledge is used in *IR*. This knowledge is presented where it is required; e.g. Logical Imaging, Situation Theory, Decision Making, Genetic Algorithms.

# 1.      LOGICS

## 1.1      Mathematical Logic (Classical First Order Logic)

Mathematical (formal, symbolic) logic is an interdisciplinary field at the intersection of mathematics and logic. It is concerned with the formal study of human thinking without taking into account the meaning of the symbols used.

Mathematical logic has two parts: *propositional calculus* and *predicate calculus*.

### 1.1.1      Propositional Calculus

#### 1.1.1.1      Proposition

The *proposition* is a statement which is known to be either *true* ($T$) or *false* ($F$).

EXAMPLE 2.1
'I am reading this text.' is a proposition because it is true. But is 'The cooks at the North Pole, wearing red hats now, are playing football.' a proposition?

There is no third alternative (principle of *tertium non datur*). $T$ and $F$ are called *truth values*.

The propositions are usually denoted by capital letters, e.g. $P, Q, R$.

They can be linked using *logical connectives* (*operators*) such as negation, disjunction, conjunction, implication, etc. to form new propositions or logical expressions whose truth–values are given by precise rules.

#### 1.1.1.2      Negation

The *negation* of a proposition $P$ is a proposition denoted by

$$\sim P \text{ (or } \neg P) \tag{2.1}$$

and pronounced 'not $P$'. If $P$ is true then $\sim P$ is false and conversely.

EXAMPLE 2.2
Let $P$ = 'The sun shines.' then $\neg P$ = 'The sun does not shine.'

### 1.1.1.3   Conjunction

The proposition '$P$ and $Q$' is denoted by

$$P \wedge Q \text{ (or } P \And Q) \tag{2.2}$$

and called a *conjunction* (logical AND). The conjunction is true if and only if both $P$ and $Q$ are true, and false otherwise.

EXAMPLE 2.3
Let $P$ = 'I am reading this text.' and $Q$ = 'Two times two is four.' then $P \wedge Q$ is true.

Notice that the conjunction may connect things which would not normally be mentioned together.

### 1.1.1.4   Disjunction

The proposition '$P$ or $Q$' is denoted by

$$P \vee Q \tag{2.3}$$

and called a *disjunction* (logical OR). The disjunction is false if and only if both $P$ and $Q$ are false, and true otherwise.

EXAMPLE 2.4
Let $P$ = 'I am reading this text.' and $Q$ = 'Two times two is five.' then $P \vee Q$ is true.

Notice that the disjunction may be true as a whole in spite of the evident falsity of one of its members.

In everyday language, the word 'or' is used in an *exclusive* sense meaning 'either ... or'; this latter meaning is expressed by a separate logical operator called *exclusive* OR (XOR) which is true if and only if $P$ and $Q$ have different truth values.

### 1.1.1.5   Implication (subjunction, conditional)

The proposition 'if $P$ then $Q$' is denoted by

$$P \Rightarrow Q \text{ (or } P \rightarrow Q) \tag{2.4}$$

and called an *implication*. The implication is false if and only if $P$ is true and $Q$ is false, and true otherwise.

EXAMPLE 2.5
Let $P$ = 'I am not reading this text.' and $Q$ = 'This text exists.' then $P \Rightarrow Q$ is true (although it may seem strange that something false implies something true).

### 1.1.1.6    Equivalence (bi–conditional)

The new proposition '$P$ is equivalent to $Q$' is denoted by

$$P \Leftrightarrow Q \text{ (other symbols used: } \equiv, \leftrightarrow) \tag{2.5}$$

and called an *equivalence*. The equivalence is true if and only if both $P$ and $Q$ have the same truth values.

EXAMPLE 2.6
Let $P$ = 'I am reading this text.' and $Q$ = 'Two times two is four.' then $P \Leftrightarrow Q$ is true. Notice, again, that the truth (or falsity) of the equivalence reflects a global truth value regardless of the particular meaning of its constituents.

### 1.1.1.7    Contradiction

The proposition

$$P \wedge \sim P \tag{2.6}$$

is called a *contradiction*. The contradiction is always false.

EXAMPLE 2.7
'The sun shines.' and 'The sun does not shine.'.

### 1.1.1.8    Tautology

A *tautology* is a proposition which is always true regardless of the truth–values of its constituent propositions. Examples:

$$P \wedge (P \Rightarrow Q) \Rightarrow Q \tag{2.7}$$

$$P \wedge Q \Rightarrow P \tag{2.8}$$

$$P \Rightarrow P \vee Q \tag{2.9}$$

$$P \Rightarrow P \tag{2.10}$$

(De Morgan's laws)  (2.11)

$$\sim(P \wedge Q) \Leftrightarrow \sim P \vee \sim Q$$
$$\sim(P \vee Q) \Leftrightarrow \sim P \wedge \sim Q$$

(idempotent laws)  (2.12)

$$P \wedge P \Leftrightarrow P$$
$$P \vee P \Leftrightarrow P$$

(involution law)  (2.13)

$$\sim\sim P \Leftrightarrow P$$

### 1.1.1.9 Rules of Inference

The general form of a *rule of inference* is as follows

$$\frac{\alpha}{\beta} \quad \text{or} \quad \alpha \vdash \beta$$

and interpreted to mean that on the assumption that $\alpha$ is true, it can be deduced that $\beta$ is true. $\alpha$ is called the *premise* and $\beta$ the *conclusion*. More important rules of inference are:

*Modus Ponens*  (2.14)

$$\frac{\begin{array}{c} P \\ P \Rightarrow Q \end{array}}{Q}$$

or $(P \wedge (P \Rightarrow Q)) \vdash Q$.

*Modus Tollens*                                                           (2.15)

$$P \Rightarrow Q$$
$$\sim Q$$

$$\overline{\phantom{xxxxx}}$$

$$\sim P$$

or $((P \Rightarrow Q) \wedge \sim Q) \vdash \sim P$.

*Reduction ad Absurdum*                                                   (2.16)

$$\sim P \Rightarrow Q$$
$$\sim P \Rightarrow \sim Q$$

$$\overline{\phantom{xxxxx}}$$

$$P$$

or $((\sim P \Rightarrow Q) \wedge (\sim P \Rightarrow \sim Q)) \vdash P$.

EXAMPLE 2.8
Let us give an example for the usage of *Reductio ad Absurdum*. Let us
assume that we want to prove that

   $P$ = "There are infinitely many primes.".

We assume first that $\sim P$ is true, i.e.,

   $\sim P$ = "There are a finite number of primes: $p_1, p_2, ..., p_n$."

It is possible now to obtain a number $N$ by adding 1 to the product of these
primes, i.e.

   $N = p_1 \cdot p_2 \cdots p_n + 1$

The number $N$ is not divisible by any of $p_1, p_2, ...., p_n$ since the remainder is
always 1. Because any number is either prime or not, $N$, too, is either
1. prime
   or
2. not.
   Let us analyse these two cases.
   If $N$ is prime, we contradict $\sim P$ ($N$ being larger than any of our primes).
It is known that any whole number can be uniquely factorised (*Fundamental
Theorem of Arithmetic*), i.e., written as a product of primes raised to powers;

these primes are the divisors of that number. If $N$ is not prime, all of its divisors are different from any of $p_1, p_2, ..., p_n$.

Because we obtain a contradiction by assuming that $\sim P$ is true, we must conclude that $P$ is true, i.e., the number of primes is infinite.

### 1.1.1.10 Normal form

The logical expression

$$A_1 \lozenge A_2 \lozenge ... A_i \lozenge ... \lozenge A_n \tag{2.17}$$

$$A_i = X_1 \square X_2 \square ... \square X_{m_i}, \quad i = 1, 2, ..., n$$

is called a *disjunctive normal form* if $\lozenge = \vee$ and $\square = \wedge$, and a *conjunctive normal form* if $\lozenge = \wedge$ and $\square = \vee$.

It can be shown that:

**THEOREM 2.1**
The implication $P \Rightarrow Q$ is equivalent to the disjunction $\sim P \vee Q$. ♦

**THEOREM 2.2**
Any logical expression can be transformed into an equivalent conjunctive (or disjunctive) normal form. ♦

### 1.1.2    Predicate Calculus

### 1.1.2.1    Predicate

A *predicate* (logical function) is a sentence with one or more variables. For example, $A(x, y)$ denotes a predicate named $A$ with two variables: $x$ and $y$. By replacing the variables with values from given sets, the predicate becomes a proposition.

### 1.1.2.2    Quantifiers

There are two quantifiers in predicate calculus:

- *universal quantifier*, denoted by $\forall$ meaning for every, for any;
- *existential quantifier*, denoted by $\exists$ meaning there exists (at least one), there is (at least one).

A special case of the existential quantifier is ∃! meaning 'there exists exactly one'.

$\forall x.A(x)$ should be read as follows: the proposition obtained from predicate $A$ for any value of variable $x$ (where $x$ takes on values from a given set).

$\exists x.A(x)$ should be read as follows: there exists at least one value $a$ of $x$ (from a given set) such that $A(a)$ is a proposition.

### 1.1.2.3    Properties

More properties which are important:

- $(\forall x.A(x) \text{ true}) \Rightarrow (\forall x.\sim A(x) \text{ false})$
- $(\forall x.A(x) \text{ false}) \Rightarrow (\forall x.\sim A(x) \text{ undecidable})$
- $(\exists x.A(x) \text{ true}) \Rightarrow (\exists x. \sim A(x) \text{ undecidable})$
- $(\exists x.A(x) \text{ false}) \Rightarrow (\exists x. \sim A(x) \text{ true})$
- $(\sim(\exists x. A(x))) \Leftrightarrow (\forall x.\sim A(x))$
- $(\sim(\forall x.A(x))) \Leftrightarrow (\exists x.\sim A(x))$
- $(\forall x.A(x)) \Rightarrow (\exists x.A(x))$
- $(((\forall x.A(x)) \wedge (\forall x.B(x))) \Leftrightarrow (\forall x.(A(x) \wedge B(x)))$
- $(((\forall x.A(x)) \vee (\forall x.B(x))) \Rightarrow (\forall x.(A(x) \vee B(x)))$
- $(((\exists x.A(x)) \vee (\exists x.B(x))) \Leftrightarrow (\exists x.(A(x) \vee B(x)))$
- $(\exists x.(A(x) \wedge B(x))) \Rightarrow ((\exists x.A(x)) \wedge (\exists x.B(x)))$

## 1.2    Non–conventional Logics

### 1.2.1    Modal Logic

*Modal Logic* is obtained by extending the Predicate Calculus with the following two operators:

- L meaning "it is necessary that", called *necessity operator*
- M meaning "it is possible that", called *possibility operator*.

More important rules of inference:

- $LA \Rightarrow A$
- $L(A \Rightarrow B) \Rightarrow (LA \Rightarrow LB)$
- $A \Rightarrow LA$
- $MA \Rightarrow LMA$

## 1.2.2    Temporal Logic

The truth values depend on time. The following operators are defined:

- F meaning "will be true at some point in the future",
- P meaning "was true at some point in the past",
- G meaning "will be true at all points in the future",
- H meaning "was true at all points in the past".

Let

$t$ denote time,
$R$ denote time precedence, i.e., $R(t, t') = t < t'$

Thus, the operators can be defined as follows ($t$ denotes present time):

- F: $\exists t'$ such that $R(t, t') \wedge A(t')$ true,
- P: $\exists t'$ such that $R(t', t) \wedge A(t')$ true,
- G $\Leftrightarrow \forall t'$F,
- H $\Leftrightarrow \forall t'$P.

## 1.2.3    Three–valued Logics

In *Kleene's Three–valued Logic*, there are three truth values as follows:

- T = true,
- F = false,
- U = undecided.

The truth value U is meant to express a state of ignorance: a variable is assigned the truth value U if it is not known to be either true or false. Negation, conjunction, disjunction, implication, and equivalence are defined as follows (Tables 2.1—2.6).

*Table 2.1* Negation in Kleene's Logic

| $\neg$A | |
|---|---|
| T | F |
| F | T |
| U | U |

Table 2.2  Conjunction in Kleene's Logic

| A ∧ B | T | F | U |
|-------|---|---|---|
| T | T | F | U |
| F | F | F | F |
| U | U | F | U |

Table 2.3  Disjunction in Kleene's Logic

| A ∨ B | T | F | U |
|-------|---|---|---|
| T | T | T | T |
| F | T | F | U |
| U | T | U | U |

Table 2.4  Implication in Kleene's Logic

| A → B | T | F | U |
|-------|---|---|---|
| T | T | F | U |
| F | T | T | T |
| U | T | U | U |

Table 2.6  Equivalence in Kleene's Logic

| A ↔ B | T | F | U |
|-------|---|---|---|
| T | T | F | U |
| F | F | T | U |
| U | U | U | U |

In *Lukasiewicz's Three–valued Logic* there are three truth vales:

–  T = true,
–  F = false,
–  I = indeterminate.

The truth value I is meant to express that a variable cannot be assigned the value true or false. The only formal difference relative to Kleene's Logic appears at implication and equivalence (Tables 2.7—2.8).

Table 2.7  Implication in Lukasievicz's Logic

| A → B | T | F | I |
|-------|---|---|---|
| T | T | F | I |
| F | T | T | T |
| I | T | I | T |

*Table 2.8* Equivalence in Lukasievicz's Logic

| A $\leftrightarrow$ B | T | F | I |
|---|---|---|---|
| T | T | F | I |
| F | F | T | I |
| I | I | I | T |

# 2. SET THEORY

## 2.1 The Concept of a Set

The concept of a set does not have a definition in a mathematical sense, rather a description.

A *set* is a collection of distinct objects. The objects in a set are called *elements*. If an object $x$ is an element of a set $S$ (or $x$ belongs to $S$), this is denoted as $x \in S$. $x \notin S$ means that $x$ does not belong to $S$. An element can occur at most once in a set. The order of the elements within a set is unimportant.

A set can be given by enumerating its elements between brackets, e.g.,

$$A = \{a_1, a_2, ..., a_n\}$$

or by giving a property (e.g., using a predicate $P$) all elements must share, e.g.

$$A = \{x|P(x)\}$$

A set having a fixed number of elements is *finite*, and *infinite* otherwise.

The *empty set* contains no elements and is denoted by $\varnothing$.

## 2.2 Subset

If all elements of a set $B$ are members of a set $A$, then $B$ is called a *subset* of $A$ and denoted by $B \subseteq A$, i.e.,

$$B \subseteq A \Leftrightarrow (\forall x \in B \Rightarrow x \in A) \tag{2.18}$$

$B \subset A$ denotes that $B$ is a *proper subset* of $A$, i.e. all elements of $B$ belong to $A$ but $A$ has other elements, too, as follows:

$$B \subset A \Leftrightarrow ((\forall x \in B \Rightarrow x \in A) \wedge (\exists y \in A \Rightarrow y \notin B)) \tag{2.19}$$

EXAMPLE 2.9
Let $B = \{1, 2, 3\}$ and $A = \{0, 1, 2, 3, 4\}$ then $B \subset A$.

## 2.3      Equality of Sets

The *equality* of two sets, $A$ and $B$, is denoted by the symbol '=' and defined
as follows:

$$A = B \Leftrightarrow ((A \subset B) \wedge (B \subset A)) \qquad\qquad (2.20)$$

i.e., $A$ and $B$ have exactly the same elements.

EXAMPLE 2.10
$A = B = \{\text{blue, red}\}$.

## 2.4      Union

The *union* of two sets, $A$ and $B$, is denoted by the symbol $\cup$ and defined as
follows:

$$A \cup B = \{x \mid x \in A \vee x \in B\} \qquad\qquad (2.21)$$

EXAMPLE 2.11
Let $B = \{1, 2, 3\}$ and $A = \{0, 1, 2, 3, 4\}$. $A \cup B = \{1, 2, 3\} \cup \{0, 1, 2, 3, 4\}$
$= \{0, 1, 2, 3, 4\}$.

## 2.5      Intersection

The *intersection* of two sets, $A$ and $B$, is denoted by the symbol $\cap$ and
defined as follows:

$$A \cap B = \{x \mid x \in A \wedge x \in B\} \qquad\qquad (2.22)$$

EXAMPLE 2.12
Let $B = \{1, 2, 3\}$ and $A = \{0, 1, 2, 3, 4\}$. $A \cap B = \{1, 2, 3\} \cap \{0, 1, 2, 3, 4\}$
$= \{1, 2, 3\}$.

If $A \cap B = \varnothing$ the sets are said to be *disjoint* sets.

## 2.6     Difference

The *difference* of two sets, $A$ and $B$, is denoted by the symbol \ and defined as follows:

$$A \setminus B = \{x \mid x \in A \wedge x \notin B\} \tag{2.23}$$

EXAMPLE 2.13
Let $B = \{1, 2, 3\}$ and $A = \{0, 1, 2, 3, 4\}$. $A \setminus B = \{0, 1, 2, 3, 4\} \setminus \{1, 2, 3\} = \{0, 4\}$.

## 2.7     Symmetrical Difference

The *symmetrical difference* of two sets, $A$ and $B$, is denoted by the symbol $\Delta$ and defined as follows:

$$A \Delta B = (A \setminus B) \cup (B \setminus A) \tag{2.24}$$

EXAMPLE 2.14
Let $B = \{1, 2, 3\}$ and $A = \{0, 1, 2, 3, 4\}$. $A \Delta B = \{0, 4\} \cup \varnothing = \{0, 4\}$.

## 2.8     Cartesian Product

The *Cartesian product* of two sets, $A$ and $B$, is denoted by the symbol $\times$ and defined as follows:

$$A \times B = \{(a, b) \mid a \in A \wedge b \in B\} \tag{2.25}$$

EXAMPLE 2.15
{red, blue} $\times$ {red, white} = {(red, red), (red, white), (blue, red), (blue, white)}.

## 2.9     Complement

Given $A \subset B$. The *complement* of set $A$ relative to set $B$ is as follows:

$$C_B A = \{x \mid x \in B \wedge x \notin A\} \tag{2.26}$$

EXAMPLE 2.16
Let $A = \{1, 2, 3\}$ and $B = \{0, 1, 2, 3, 4\}$. $C_B A = \{0, 4\}$.

## 2.10    Power Set

The *power set* of $A$ is as follows:

$$\wp(A) = \{X \mid X \subseteq A\} \tag{2.27}$$

The empty set is always a member of the power set.

EXAMPLE 2.17
$\wp(\{\text{red, blue}\}) = \{\varnothing, \{\text{red}\}, \{\text{blue}\}, \{\text{red, blue}\}\}$.

## 2.11    Cardinality

The concept of *cardinality* is the expression of the common property of having the same number of elements.

The *cardinality* of a set $A$ is denoted by $|A|$ and defined as the number of elements it contains. The cardinality of a finite set $A$ is equal to a positive integer $n$, i.e., $|A| = n$, whereas that of an infinite set $B$ is equal to infinity, i.e., $|B| = +\infty$.

The cardinality of the power set $\wp(A)$ is $2^n$ where $|A| = n$.

EXAMPLE 2.18
$|\wp(\{\text{red, blue}\})| = 2^2 = 4$. $|\mathbf{N}| = +\infty$ (where $\mathbf{N}$ denotes the set of natural numbers).

## 2.12    Sets of Numbers

The following sets of numbers are commonly used in *IR*:

$\mathbf{N}$   the set of natural numbers,
$\mathbf{Z}$   the set of integer numbers,
$\mathbf{R}$   the set of real numbers.

# 3.     RELATIONS

## 3.1     Binary Relation

Given two sets $A$ and $B$. A *binary relation* $R$ is a subset of the Cartesian product $A \times B$, i.e.

$$R \subseteq A \times B \tag{2.28}$$

$A$ is the *domain*, $B$ is the *codomain* of the relation. The fact that $(x, y) \in R$ can also be denoted as follows: $x R y$, and should be read as $x$ is in relation $R$ with $y$.

EXAMPLE 2.19
Let $A = \{1, 2, 3\}$ and $B = \{0, 1, 2, 3, 4\}$. $R = \{(1, 1), (2, 2), (3, 3)\} \subseteq A \times B$ is a binary relation (namely, equality).

## 3.2     Equivalence Relation

A binary relation $R \subseteq A \times A$ is an *equivalence relation* if it is:

- reflexive: $x R x, \forall x \in A$;
- symmetric: $((x, y \in A) \wedge (x R y)) \Rightarrow y R x$;
- transitive: $((x, y, z \in A) \wedge (x R y) \wedge (y R z)) \Rightarrow x R z$.

EXAMPLE 2.20
The binary relation = (equality) over $\mathbf{N}$ is an equivalence relation.

## 3.3     Ordering Relation

A binary relation $R \subseteq A \times A$ is an *ordering relation* if it is:

- reflexive: $x R x, \forall x \in A$;
- transitive: $((x, y, z \in A) \wedge (x R y) \wedge (y R z)) \Rightarrow x R z$;
- antisymmetric: $((x, y \in A) \wedge (x R y) \wedge (y R x)) \Rightarrow x = y$.

EXAMPLE 2.21
The binary relation $\leq$ (less than or equal to) over $\mathbf{N}$ is an ordering relation.

It can be shown that:

**THEOREM 2.3**
Given a set $A$. There is exactly one relation on A that is reflexive, symmetric and antisymmetric, namely the identity relation (the elements in any pair are equal to each other). ♦

## 3.4     Partially Ordered Set (Poset)

A set with an ordering relation defined on it is a *partially ordered set* (or *poset*).

EXAMPLE 2.22
N of Example 2.21 is a poset.

## 3.5     Partition

The partition of a set $A$ is given by mutually disjoint subsets $X_1, X_2, ..., X_n$ as follows:

$$A = X_1 \cup X_2 \cup ... \cup X_n, \tag{2.29}$$

$$X_i \cap X_j = \varnothing, i = 1, 2, ..., n, j = 1, 2, ..., n, i \neq j$$

EXAMPLE 2.23
$X_1 = \{0\}, X_2 = \{1, 2\}$ and $X_3 = \{3, 4\}$ is a partition of the set $\{0, 1, 2, 3, 4\}$.

It can be shown that:

**THEOREM 2.4**
An equivalence relation on $A$ provides a partitioning of $A$ into mutually disjoint equivalence classes. ♦

## 4.     FUNCTION

## 4.1     The Concept of a Function

Given a binary relation

$$f = \{(x, y) \mid x \in A, y \in B\} \subseteq A \times B$$

$f$ is called a *function* (or mapping) if $\forall x \in A \; \exists!$ (there exists exactly one) $y \in B$ such that $(x, y) \in f$.

Usually a function $f$ from $A$ into $B$ is denoted as follows:

$$f: A \to B, \; y = f(x),$$

$A$ is the *domain*, $B$ is the *codomain*, $y$ is the *value of function $f$* at point $x$ or the *image* of $x$ under $f$.

EXAMPLE 2.24

$f: \mathbf{N} \to \mathbf{N}, f(x) = 2x, \; f: \mathbf{R} \to \mathbf{R}, f(x) = x^3$.

## 4.2   Injection

A function $f: A \to B$ is an *injection* if different elements are mapped onto different elements, i.e.,

$$x_1 \neq x_2 \Rightarrow f(x_1) \neq f(x_2) \tag{2.30}$$

or, equivalently, $f(x_1) = f(x_2) \Rightarrow x_1 = x_2$.

EXAMPLE 2.25

$f: \mathbf{N} \to \mathbf{N}, f(x) = x^3$ is an injection whilst $f: \mathbf{Z} \to \mathbf{N}, f(x) = x^2$ is not (because $f(-2) = f(2) = 4$).

It can be shown that:

**THEOREM 2.5**

If the function $f: A \to B$ is injective and $X \subseteq A$ then $f(C_A(X)) = C_{f(A)}(f(X))$ ♦

## 4.3   Surjection

A function $f: A \to B$ is a *surjection* if every element of the codomain is an image of some element of the domain, i.e.

$$\forall y \in B \; \exists x \in A \text{ such that } y = f(x) \tag{2.31}$$

EXAMPLE 2.26

$f: \mathbf{N} \to \mathbf{N}, f(x) = x$ is a surjection whilst $f: \mathbf{N} \to \mathbf{N}, f(x) = 2x$ is not (because e.g., 3 is not the double of any natural number).

## 4.4     Bijection and Inverse

A function $f: A \to B$ that is both an injection and a surjection is a *bijection*. The domain and codomain of a bijection have the same cardinality. There is a unique function $f^{-1}: B \to A$ and called the *inverse* function of $f$.

EXAMPLE 2.27
$f: \mathbf{R} \to \mathbf{R}, f(x) = 2x + 1$ is a bijection and its inverse is the function $f^{-1}(y) = (y - 1)/2$

## 4.5     Restriction of a Function

The *restriction* $f|_C$ of function $f: A \to B$ is a new function whose domain $C$ is a subset of $A$, i.e. $C \subseteq A$.

EXAMPLE 2.28
The restriction of $f: \mathbf{R} \to \mathbf{R}, f(x) = 2x + 1$ to $\mathbf{N}$ is $f|_\mathbf{N}: \mathbf{N} \to \mathbf{R}$ (which is no longer a bijection).

## 4.6     Composition of Functions

Given two functions $f: A \to B$ and $g: B \to C$. The function

$$h: A \to C, h(x) = g(f(x)) \tag{2.32}$$

is the *composition* of $f$ and $g$, and is denoted by $g \circ f$.

EXAMPLE 2.29
The composition of $f: \mathbf{N} \to \mathbf{N}, f(x) = 2x + 1$ and $g: \mathbf{N} \to \mathbf{N}, g(x) = x^2$ is $h(x) = g(f(x)) = (2x + 1)^2$.

It can be shown that:

THEOREM 2.6
If the functions $f: A \to B$ and $g: B \to C$ are surjective then so is $g \circ f$. ♦

THEOREM 2.7
If the functions $f: A \to B$ and $g: B \to C$ are injective then so is $g \circ f$. ♦

**THEOREM 2.8**

If the composition $g \circ f$ of the functions $f: A \to B$ and $g: B \to C$ is surjective then so is $g$. ♦

**THEOREM 2.9**

If the composition $g \circ f$ of the functions $f: A \to B$ and $g: B \to C$ is injective then so is $f$. ♦

## 4.7    Fixed Point

Given a function $f: A \to B$. The point $x$ is a *fixed point* of $f$ if $x = f(x)$.

EXAMPLE 2.30

The fixed point of $f: \mathbf{N} \to \mathbf{N}$, $f(x) = x^2$ is $x = 1$ because $f(1) = 1$ (1 is the solution of the equation $x = x^2$ in $\mathbf{N}$).

It can be shown that:

**THEOREM 2.10**

Given $f: A \to A$. Let $f^n: A \to A$ denote the function defined recursively by $f^n = f \circ f^{n-1}$. If there exists $n \in \mathbf{N}$, $n \neq 0$, such that $f^n$ has exactly one fixed point, then $f$ has at least one fixed point. ♦

## 4.8    Homeomorphism

A (continuous) function $g: A \to B$ is called a *homeomorphism* if its inverse function $g^{-1}: B \to A$ exists and is continuous; $g^{-1}(g(a)) = a$, $\forall a \in A$, $g(g^{-1}(b)) = b$, $\forall b \in B$ ($A$, $B$ usually are topological spaces).

## 5.    FAMILY OF SETS

## 5.1    The Concept of a Family of Sets

Given a function

$$f: I \to \wp(A), \quad f(i) = A_i \subseteq A, \quad \forall i \in I$$

Then $f$ defines a *family of sets* on $A$; it is denoted by $(A_i)_{i \in I}$.

## 5.2      Union

The *union* of a family of sets $(A_i)_{i \in I}$ is defined as the set of all elements of $A$ that belong to at least one $A_i$:

$$\bigcup_{i \in I} A_i = \{x \mid x \in A \wedge \exists i \in I \text{ such that } x \in A_i\} \qquad (2.33)$$

## 5.3      Intersection

The *intersection* of a family of sets $(A_i)_{i \in I}$ is defined as the set of all elements of $A$ that belong to every $A_i$:

$$\bigcap_{i \in I} A_i = \{x \mid x \in A \wedge x \in A_i, \forall i \in I\} \qquad (2.34)$$

It can be shown that:

**THEOREM 2.11**
Given a family of sets $(A_i)_{i \in I}$, $A_i \subseteq A$, $\forall i \in I$, and let $g: A \to B$ be a function. Then $g(\bigcap_{i \in I} A_i) \subseteq \bigcap_{i \in I} g(A_i)$ and $g(\bigcup_{i \in I} A_i) = \bigcup_{i \in I} g(A_i)$. ♦

## 5.4      Covering

The family of sets $(A_i)_{i \in I}$ is a *covering* of set $B \subseteq A$ if $B \subseteq \bigcup_{i \in I} A_i$.

## 5.5      Mutually Disjoint Family of Sets

The family of sets $(A_i)_{i \in I}$ is *mutually disjoint* if

$$A_i \cap A_j = \emptyset, \quad \forall i, j \in I, \ i \neq j \qquad (2.35)$$

## 5.6      Partition

Given a nonempty set $A$, i.e., $A \neq \emptyset$. The family of sets $(A_i)_{i \in I}$ is a *partition* of $A$ if $(A_i)_{i \in I}$ is a mutually disjoint covering of $A$ and $A_i \neq \emptyset$, $\forall i \in I$.

# 6. ALGEBRA

## 6.1 Elementary Functions

### 6.1.1 Exponential Function

The *exponential function* is defined as follows:

$$f: \mathbf{R} \to (0, +\infty), \quad f(x) = a^x, \quad a \in \mathbf{R} \text{ constant, } a > 0, \ a \neq 1$$

It can be shown that the exponential function is bijective, its inverse function is the logarithmic function (see below), and it is strictly increasing if $a > 1$, and strictly decreasing if $a < 1$.

### 6.1.2 Logarithmic Function

The *logarithmic function* is defined as follows:

$$f: (0, +\infty) \to \mathbf{R}, \quad f(x) = \log_a x, \quad a \in \mathbf{R} \text{ constant, } a > 0, \ a \neq 1$$

It can be shown that the logarithmic function is bijective, its inverse function is the exponential function (see above), and it is strictly increasing if $a > 1$, and strictly decreasing if $a < 1$. a is called the base of the logarithm.
   There are three most commonly used logarithms:

- base 10, $a = 10$, (common or Briggs system), notation: $\log_{10}$
- base 2, $a = 2$, notation: $\log_2$
- natural (or Naperian) logarithm, $a = e = 2.7182818284...$ (Euler's number), notation: ln

It can be shown that the following important properties hold:

- $\log_a(x/y) = \log_a x - \log_a y$
- $\log_a(xy) = \log_a x + \log_a y$
- $\log_a(x^y) = y \cdot \log_a x$
- $\log_a a = 1$
- $\log_a 1 = 0$
- $(\log_a b) \cdot (\log_b a) = 1$
- $\log_a x = (\log_a b) \cdot (\log_b x)$

## 6.2      Permutation

A bijection of a set $S$ onto itself is a called a *permutation*. When $S$ is finite a permutation of $S$ means a rearrangement of its elements. The number $P_n$ of permutations (without repetition) of $n$ elements, i.e., $|S| = n$, is

$$P_n = 1 \cdot 2 \cdot \ldots \cdot n = n! \tag{2.36}$$

$n!$ denotes the *factorial* of $n$.

EXAMPLE 2.31
$P_3 = 6$: 1, 2, 3; 1, 3, 2; 2, 1, 3; 2, 3, 1; 3, 1, 2; 3, 2, 1.

## 6.3      Combination

A subset $K$ of a set $S$, $K \subseteq S$, $|K| = k$, $|S| = n$, is called a *combination*. The number $^nC_k$ of combinations of $n$ elements taken $k$ at a time is

$$^nC_k = \frac{n!}{(k!(n-k)!)} \tag{2.37}$$

It can be shown that the following more important properties hold:

- $^nC_k = {}^nC_{n-k}$
- $^nC_0 = {}^nC_n = {}^0C_0 = 1$
- $^nC_1 = n$

Newton's *binomial formula* is as follows:

$$(a+b)^N = \sum_{k=1}^{N} {}^NC_k a^{N-k} b^k \tag{2.38}$$

It follows that:

$$\sum_{k=1}^{N} {}^NC_k = 2^N \tag{2.39}$$

## 6.4 More Important Inequalities

### 6.4.1 Bernoulli's Inequality

$$(1 + a)^n \geq 1 + na, \quad a \in (-1, +\infty), \quad n \in \mathbf{N} \tag{2.40}$$

### 6.4.2 Cauchy–Bunyakowsky–Schwarz's Inequality

Given $\forall a_i, b_i \in \mathbf{R}, \quad i = 1, 2, ..., n$:

$$\left( \sum_{i=1}^{n} a_i b_i \right)^2 \leq \left( \sum_{i=1}^{n} a_i^2 \right) \left( \sum_{i=1}^{n} b_i^2 \right)$$

$$\tag{2.41}$$

### 6.4.3 Hölder's Inequality

Given $q, p, a_i, b_i \in \mathbf{R}_+ = (0, +\infty), \quad i = 1, 2, ..., n, \quad 1/p + 1/q = 1$:

if $p, q > 1$:

$$\sum_{i=1}^{n} a_i b_i \leq \sqrt[p]{\sum_{i=1}^{n} a_i^p} \cdot \sqrt[q]{\sum_{i=1}^{n} b_i^q}$$

$$\tag{2.42}$$

if $0 < p < 1, \quad 0 < q < 1$:

$$\sum_{i=1}^{n} a_i b_i \geq \sqrt[p]{\sum_{i=1}^{n} a_i^p} \cdot \sqrt[q]{\sum_{i=1}^{n} b_i^q}$$

$$\tag{2.43}$$

## 6.5 Matrices

Given elements $a_{ij}$. In $IR$ these are typically real numbers such as weights, probabilities. Therefore we assume that $a_{ij} \in \mathbf{R}, \quad i = 1, 2, ..., m, \quad j = 1, 2, ..., n$.

The table

$$\begin{bmatrix} a_{11} & a_{12} & ... & a_{1n} \\ a_{21} & a_{22} & ... & a_{2n} \\ ... ... ... & a_{ij} & ...... \\ a_{m1} & a_{m2} & ... & a_{mn} \end{bmatrix}$$

is called a *matrix*. The numbers $a_{ij}$ are called the *elements* of the matrix. The name of the matrix is usually a capital letter, e.g., $A$. It has $m$ *rows* and $n$ *columns*. Other notations are $A_{m,n}$, $(a_{ij})_{m,n}$ The matrix $A_{m,1}$ is called a *column vector*, and $A_{1,n}$ a *row vector*. An arbitrary element is denoted by $a_{ij}$, i.e. the element on row $i$ and column $j$. In what follows, real matrices will be considered only. The elements $a_{ii}$ form the *main diagonal* of matrix $A$. A matrix is a *square matrix* if the number of its rows and columns is the same, i.e. $A_{n,n}$. If all elements, apart from those on the main diagonal, are null the matrix is called a *diagonal* matrix; notation: diag($A$).

### 6.5.1 Unit (Identity) Matrix

The *unit* matrix $I_{n,n}$ is defined as the matrix whose elements are all equal to 0 apart from those on the main diagonal which are all equal to 1.

EXAMPLE 2.32
$I_{3,3}$ is the following matrix:

$$I_{3,3} = \begin{bmatrix} 1 & 0 & 0 \\ 0 & 1 & 0 \\ 0 & 0 & 1 \end{bmatrix}$$

### 6.5.2 Addition

Given two matrices $A_{m,n}$ and $B_{m,n}$. Their *sum* matrix $C$ is

$$C = (c_{ij})_{m,n} = A + B$$

where

$$c_{ij} = a_{ij} + b_{ij}, \quad i = 1, 2, ..., m, \quad j = 1, 2, ..., n \tag{2.45}$$

### 6.5.3    Null Matrix

The matrix whose elements are all equal to 0 is called the *null matrix* and denoted by $0_{m,n}$.

### 6.5.4    Multiplication

Given two matrices $A_{m,n}$ and $B_{n,p}$. Their *product* matrix is

$$C = (c_{ik})_{m,p} = AB \tag{2.46}$$

where

$$c_{ik} = a_{i1}b_{1k} + a_{i2}b_{2k} + ... + a_{in}b_{nk}, \quad i = 1, 2, ..., m, \quad k = 1, 2, ..., p \tag{2.47}$$

It can be shown that the addition and multiplication of matrices have the following properties:

- $A + B = B + A \quad \forall A, B$ (commutativity)
- $(A + B) + C = A + (B + C), \forall A, B, C$ (associativity)
- $A_{m,n} + 0_{m,n} = 0_{m,n} + A = A$
- $A + (-A) = (-A) + A = 0_{m,n}$
- $(AB)C = A(BC)$ (associativity)
- $AB \neq BA$ (not commutative although there are exeptions)
- $A(B + C) = AB + AC$
- $(A + B)C = AC + AC$
- $AI_{n,n} = I_{n,n} A = A$

### 6.5.5    Multiplication by a Scalar

Given a matrix $A_{m,n}$ and a real number (referred to as a *scalar*) $k \in \mathbf{R}$. The multiplication of matrix $A$ by the scalar $k$ is

$$B = (b_{ij})_{m,n} = kA \tag{2.48}$$

where

$$b_{ij} = ka_{ij}, \quad i = 1, 2, ..., m, \quad j = 1, 2, ..., n \tag{2.49}$$

It can be shown that the following properties hold ($\forall a, b \in \mathbf{R}, \forall A_{m,n}, B_{m,n}$):

- $(a + b)A = aA + bA$
- $a(A + B) = aA + aB$
- $a(bA) = (ab)A$
- $a(AB) = (aA)B$

### 6.5.6    Transpose

The *transpose* of matrix $A_{m,n}$ is the matrix $A^T = (a_{ji})_{n,m}$. It can be shown that the following properties hold:

- $(A^T)^T = A$
- $(kA)^T = kA^T$
- $(AB)^T = B^T A^T$
- $(A + B)^T = A^T + B^T$

### 6.5.7    Symmetric Matrix

The matrix $A_{n,n}$ is *symmetric* if $a_{ij} = a_{ji}$, $i = 1, 2, ..., n$, $j = 1, 2, ..., n$.

### 6.5.8    Skew–symmetric Matrix

The matrix $A_{n,n}$ is *skew symmetric* if $a_{ij} = -a_{ji}$, $i = 1, 2, ..., n$, $j = 1, 2, ..., n$.

### 6.5.9    Orthogonal Matrix

A matrix $A_{m,n}$ is *orthogonal* if $AA^T = I_{m,m}$.

## 6.6    Determinant

Given a square matrix $A_{n,n}$. The *determinant of order n* of matrix $A$ is a real number associated to $A$, denoted by $det(A)$ or $|A|$ and calculated as

$$det(A) = \sum (-1)^J a_{1 j_1} a_{2 j_2} \cdots a_{n j_n}$$

$$(2.50)$$

where the sum is taken over all $n!$ permutations $j_1, j_2, ..., j_n$, and $J$ is the parity of a permutation.

For large determinants there are special numerical procedures to calculate them.

A determinant can also be represented as the elements of the matrix between vertical lines.

It can shown that the value of the determinant is null if:
— all elements on a row (column) are null,
— two rows (columns) are equal or proportional.

### 6.6.1 Rank

The *rank* of a matrix $A_{m,n}$ is the order of the largest non–zero determinant that can be obtained from $A$. Notation: rank($A$).

### 6.6.2 Singular (Regular) Matrix

The matrix $A_{n,n}$ is *singular* if det(A) = 0. It is *regular* (non–singular) if det(A) $\neq$ 0.

### 6.6.3 Inverse Matrix

Given a regular square matrix $A_{n,n}$. Its *inverse matrix* is denoted by $A^{-1}$ and the following property holds:

$$A \cdot A^{-1} = I \tag{2.51}$$

There are special numerical procedures to compute the inverse.

## 6.7 Simultaneous System of Linear Equations

It has the form

$$A_{n,m}X_{m,1} = B_{n,1} \text{ or } A_{n,m}\mathbf{x}_{m,1} = B_{n,1} \tag{2.52}$$

where $A$ and $B$ are given, whereas $\mathbf{x}$ denotes the *unknowns* for which the system is to be solved, i.e., determine those values of $\mathbf{x}$ for which the equality holds.

EXAMPLE 2.33

$$2x_1 + 3x_2 = 1$$
$$x_1 - 5x_2 = 4$$

or using a matrix notation ($n = m = 2$)

$$
\begin{bmatrix} 2 & 3 \\ 1 & -5 \end{bmatrix} \begin{bmatrix} x_1 \\ x_2 \end{bmatrix} = \begin{bmatrix} 1 \\ 4 \end{bmatrix}
$$

$$
A_{2,2} \qquad\qquad X_{2,1} \qquad B_{2,1}
$$

There are special procedures to solve a simultaneous system of linear equations. It is not always possible to solve the system, in which case the system is said to be *incompatible*. It is said to be *compatible* if it is soluble.

Let $r$ denote the rank of $A$. It can be shown that if $r = n$ the system is compatible and can be solved. If $r < n$ the system does not always have a solution.

## 6.8    Singular Value Decomposition

### 6.8.1    Eingenvalues

Let $A_{n,n}$ be a regular matrix. The solutions (roots) of the following $n$–degree polynomial equation (called the *characteristic equation*):

$$
|A - \lambda I| = 0 \tag{2.53}
$$

are called the *eigenvalues* (*characteristic* or *latent roots*) of $A$.

EXAMPLE 2.34
Let

$$
A = \begin{bmatrix} 1 & 5 \\ 2 & 4 \end{bmatrix}
$$

be a regular matrix, $\det(A) = |A| = -6 \neq 0$. The characteristic equation is:

$$
\begin{vmatrix} 1 - \lambda & 5 \\ 2 & 4 - \lambda \end{vmatrix} = 0
$$

i.e., $(1 - \lambda)(4 - \lambda) - 5 \cdot 2 = 0$ which becomes $\lambda^2 - 5\lambda - 6 = 0$. The eigenvalues are $\lambda_1 = 6$ and $\lambda_2 = -1$.

It can be shown that:

**THEOREM 2.12**
The diagonal elements of a diagonal matrix are its eigenvalues. ◆

### 6.8.2 Eigenvectors

Let $\lambda_i$, $i = 1, ..., n$, be the eigenvalues of matrix $A_{n,n}$. The vectors (column matrix) $\mathbf{X}_i$ satisfying the simultaneous system of linear equations:

$$(A - \lambda_i I)\mathbf{X}_i = 0 \tag{2.54}$$

are called the *eigenvectors* (*characteristic* or *latent vectors*) of $A$.

It can be shown that:

**THEOREM 2.13**
Eigenvectors corresponding to distinct eigenvalues are linearly independent of each other. ◆

### 6.8.3 Canonical Form

Let the eigenvalues of $A_{n,n}$ be $\lambda_1, ..., \lambda_n$. The diagonal matrix

$$D = \text{diag}(\lambda_1, ..., \lambda_n) \tag{2.55}$$

(the order of $\lambda_i$ is arbitrary) is called the *canonical form* form of $A$. Obviously, if all eigenvalues are distinct, there are $n!$ canonical forms.

### 6.8.4 Similarity Matrix

Given a regular square matrix $A_{n,n}$ which has $n$ distinct eigenvalues $\lambda_1, ..., \lambda_n$. Let the corresponding eigenvectors be $\mathbf{X}_1, ..., \mathbf{X}_n$. The matrix $S$ defined as:

$$S = [\mathbf{X}_1, ..., \mathbf{X}_n] \tag{2.56}$$

is termed a *modal matrix* of $A$. It can be shown that:

**THEOREM 2.14**
The matrix $A$ can be reduced to a canonical form $D$ using its modal matrix $S$ as a similarity matrix, as follows: $S^{-1}AS = D$ ◆

### 6.8.5    Singular Value Decomposition

Given a matrix $A_{m,n}$, $m \geq n$, rank($A$) = $r$. The *singular value decomposition* (*SVD*) of $A_{m,n}$ is defined as follows:

$$A = UDV^T \qquad\qquad (2.57)$$

where $U^T U = V^T V = I_{n,n}$ and $D$ is a diagonal matrix $D = \text{diag}(d_1, ..., d_n)$, $d_i > 0$, $i = 1, ..., r$, $d_j = 0$, $j > r$. Matrices $U$ and $V$ are orthogonal, and their first $r$ columns define the orthonormal eigenvectors associated with the $r$ non–zero eigenvalues of $AA^T$ and $A^T A$, respectively. The columns of $U$ are called the *left singular vectors*, and those of $V$ the *right singular vectors*. The diagonal elements of $D$ are the non–negative square roots of the $n$ eigenvalues of $AA^T$, and are referred to as the *singular values* of $A$.


## 7.    CALCULUS

## 7.1    Sequence

Given a set $A$. A function

$$f\colon \mathbf{N} \to A , \quad a_n = f(n) \qquad\qquad (2.58)$$

defines a *sequence* in $A$. More usual notations are:

$a_1, a_2, ..., a_n, ...$

or

$\{a_n\}$

or simply (if there is no danger of misunderstanding)

$(a_n), a_n$

If $A = \mathbf{R}$ we have a *sequence of real numbers*. If $A$ is a set of functions we have a *sequence of functions*.

EXAMPLE 2.35
$a_n = n^2$, $n = 1, 2, ...$ defines the sequence of squares of natural numbers.

## 7.2    Limit

The sequence of real numbers $a_n$ has the *limit* $L \in \mathbf{R}$, or *tends* to $L$, if

$$\forall \varepsilon \in \mathbf{R} \; \exists n_\varepsilon \in \mathbf{N} \text{ such that } (\forall n > n_\varepsilon \Rightarrow |a_n - L| < \varepsilon) \tag{2.59}$$

A sequence that has a (finite) limit is said to *converge*; it is *divergent* otherwise. Not every sequence has a limit.
   Notations: $a_n \to L$ when $n \to +\infty$ or

$$\lim_{n \to +\infty} a_n = L$$

EXAMPLE 2.36

$$\lim_{n \to +\infty} \frac{1}{n} = 0$$

It can be shown that:

**THEOREM 2.15**
A convergent sequence has a unique limit. ♦

## 7.3    Cauchy Sequence

A sequence $a_n$ is called a *Cauchy sequence (fundamental sequence)* if

$$\forall \varepsilon \in \mathbf{R} \; \exists n_\varepsilon \in \mathbf{N} \text{ such that } ((\forall n > n_\varepsilon) \wedge (\forall m > n_\varepsilon)) \Rightarrow |a_n - a_m| < \varepsilon) \tag{2.60}$$

It can be shown that:

**THEOREM 2.16**
Every convergent sequence is a Cauchy sequence. ♦

## 7.4    Limit of Functions

A function $f$ has a *limit* $b$ at $x = a$ if for $x$ sufficiently close to $a$ the values of $f$ are sufficiently close to $b$, i.e.,

$$\lim_{x \to a} f(x) = b \Leftrightarrow$$

$$(\forall \varepsilon > 0 \; \exists \delta_\varepsilon > 0 \text{ such that } (|x - a| < \delta_\varepsilon \Rightarrow |f(x) - b| < \varepsilon)) \tag{2.61}$$

It can be shown that:

**THEOREM 2.17**
A function $f$ has a limit $b$ at $x = a$ if for any sequence $x_n$ (from the domain) that has limit $a$ the corresponding sequence of function values has limit $b$, i.e., $\forall x_n \to a \Leftrightarrow f(x_n) \to b$. ◆

## 7.5     Continuous Function

A function $f: A \to B, A, B \subseteq \mathbf{R}$, is *continuous* at point $x = a$ if it has a limit at point $a$ and $\lim_{x \to a} f(x) = f(a)$, and $f$ is *not continuous* otherwise.

EXAMPLE 2.37
The function

$$f(x) = \begin{cases} \dfrac{\sin x}{x}, x \neq 0 \\ 0, x = 0 \end{cases}$$

is not continuous at $x = 0$.

## 7.6     Derivative

Given a function $f: A \subseteq \mathbf{R} \to B \subseteq \mathbf{R}, y = f(x)$. Let $\Delta x$ denote an increment of $x$ and $\Delta y$ the corresponding increment of $y$. The *derivative* of $y$ with respect to $x$ is:

$$\lim_{\Delta x \to 0} \frac{\Delta y}{\Delta x}$$

The derivative is usually denoted by $y'$ or $dy/dx$. This is called the *first order derivative*. The first order derivative of $y'$ is called the *second order derivative* of $y$ and is denoted by $y''$, and so on.
    If $y'$ is finite $f$ is said to be *differentiable*.
    There are rules to calculate the derivatives of different functions.

EXAMPLE 2.38

| function $y = f(x)$ | first order derivative $y'$ |
|---|---|
| $C$ (constant) | 0 |
| $x^n$ | $nx^{n-1}$ |
| $\sin(x)$ | $\cos(x)$ |
| $\ln(x)$ | $1/x$ |

A function $f(x)$ is *decreasing* if $df/dx < 0$ and *increasing* if $df/dx > 0$.
It can be shown that:

**THEOREM 2.18**
If $f$ is differentiable at point $x = a$ it is also continuous at that point. The
reverse is not always true. ◆

## 7.7    Maximum and Minimum

Given a function $f: A \subseteq \mathbf{R} \to B \subseteq \mathbf{R}, y = f(x)$. The solutions of the equation

$$y' = 0$$

are the *stationary points* of $f$. A stationary point can be a *maximum* or a
*minimum*, i.e. the points at which $f$ takes maximal/minimal values. Whether
they are minima or maxima can be established by computing the value of the
second order derivative $y''$ at those points: if the value of $y''$ is positive we
have a minimum, if it is negative we have a maximum, if it is null we have
an *inflection* (*saddle point*).

## 7.8    The Indefinite Integral

The set of all functions whose derivative is equal to a given function $f$ is
called the *indefinite integral* of $f$ (or *primitive functions* of $f$) and is denoted
by

$$\int f(x)dx$$

Technically, differentiation and integration are inverse operations. The primitive functions are computed based on differentiation rules.

EXAMPLE 2.39
The primitive functions of $f(x) = 2x$ are given by

$$\int 2x dx = x^2 + c$$

where $c \in \mathbf{R}$ is an arbitrary constant (its derivative is null) and the derivative of $x^2$ is $d(x^2)/dx = 2x$.

## 7.9    The Definite Integral

Given a function $f: A \subseteq \mathbf{R} \to B \subseteq \mathbf{R}, y = f(x)$. The area bounded by the graph of $f$, the $x$ axis, and the vertical lines $x = a$ and $x = b$ is represented by the *definite integral* which is obtained by placing the values $a$ and $b$ after the integration sign as follows:

$$\int_a^b f(x) dx$$

$a$ and $b$ are called the *integration limits*. The *value* of the definite integral is a number that can be calculated using, for example, the *Newton—Leibniz formula*:

$$\int_a^b f(x) dx = F(b) - F(a)$$

where $F$ is a primitive function of $f$, i.e., $F' = f$. This represents a connection between the indefinite and definite integral.

EXAMPLE 2.40

$$\int_0^1 2x dx = (x^2)(1) - (x^2)(0) = 1$$

There also are approximate methods for calculating the value of a definite integral.

# 8.    DIFFERENTIAL EQUATIONS

## 8.1    Linear Equation of the First Degree (type 1 linear equation)

The *type* 1 *linear differential equation* is as follows:

$$y' + f(x) \cdot y + g(x) = 0 \tag{2.62}$$

where $y = y(x)$ is a real function of $x$ and is to be found, whereas $f(x)$ and $g(x)$ are given (continuous) real functions. The solution is given by:

$$y(x) = e^{-\int_{x_0}^{x} f(t)dt} \left( K - \int_{x_0}^{x} g(t) e^{\int_{x_0}^{t} f(u)du} dt \right) \tag{2.63}$$

where $e$ is the base of the natural logarithm ln and $K$ is a constant.

EXAMPLE 2.41
The solution of the equation

$$y' - \frac{y}{x} - \frac{\ln x}{x} = 0$$

is

$$y(x) = Kx - \ln x - 1$$

## 8.2    Cauchy Problem

The *Cauchy problem* (or differential equation with *initial conditions*) is defined as follows:

$$y' + f(x) \cdot y + g(x) = 0 \tag{2.64}$$

$$y(x_0) = y_0$$

where $y_0$ denotes the *initial value* of $y$ at an *initial point* $x_0$. The solution is obtained by computing the general solution first, and then calculating the value of $K$ using the initial condition $y(x_0) = y_0$.

EXAMPLE 2.42
The solution of the following Cauchy problem

$$y' - \frac{y}{x} - \frac{\ln x}{x} = 0$$

$$y(1) = -1$$

is

$$y(x) = -\ln x - 1$$

## 9.     VECTORS

A *vector* is a quantity which has a *magnitude* and a *direction*. Vectors are usually denoted by bold lowercase letters. For example, the velocity of a car is a vector, say, $v$: $v = 100$ km/h (magnitude) southwestwards (direction).
    Let

$$\mathbf{R}^n = \mathbf{R} \times \mathbf{R} \times ... \times \mathbf{R} \quad (n \text{ times})$$

be an $n$–dimensional system of axes which are mutually perpendicular to each other (*Euclidean space*), and let $O$ denote the origin of this system. Let $u_i$, $i = 1, 2, ..., n$, denote the vector of length (or magnitude) 1 (unit) situated on the $i$th axis pointing from origin $O$ outwards; $u_i$ is referred to as a *basic unit vector*. Let $v$ denote an arbitrary vector pointing from $O$ to an arbitrary direction in space, and let $v_i$ denote the length of the perpendicular projection of $v$ onto the $i$th axis, $i = 1, 2, ..., n$. It can be shown that:

$$\mathbf{v} = v_1\mathbf{u}_1 + v_2\mathbf{u}_2 + ... + v_n\mathbf{u}_n \tag{2.65}$$

This is why a vector $v$ can also be represented as

$$\mathbf{v} = (v_1, v_2, ..., v_n)$$

The length $v$ of the vector $v$ is given by

$$v = \sqrt{v_1^2 + v_2^2 + \ldots + v_n^2}$$

## 9.1 Sum of Vectors

The *sum* of vectors $\mathbf{v} = (v_1, v_2, \ldots, v_n)$ and $\mathbf{w} = (w_1, w_2, \ldots, w_n)$ is:

$$\mathbf{v} + \mathbf{w} = (v_1 + w_1, v_2 + w_2, \ldots, v_n + w_n) \tag{2.66}$$

## 9.2 Scalar Product

The *scalar (dot) product* is

$$\mathbf{v} \cdot \mathbf{w} = vw \cos \theta \tag{2.67}$$

where $v$ is the length of $\mathbf{v}$, $w$ the length of $\mathbf{w}$ and $\theta$ is the angle between the two vectors.

## 9.3 Vector (Linear) Space

Given a real vector $\mathbf{v} = (v_1, v_2, \ldots, v_n) \in \mathbf{R}^n$ and a scalar $k \in \mathbf{R}$. The product $k\mathbf{v}$ is called the multiplication of $\mathbf{v}$ by scalar $k$ and is defined as follows:

$$k\mathbf{v} = (kv_1, kv_2, \ldots, kv_n) \tag{2.68}$$

The addition of vectors and multiplication by scalars satisfy the following properties:

- $\mathbf{v} + \mathbf{w} = \mathbf{w} + \mathbf{v}$ (commutativity) $\forall \mathbf{v}, \mathbf{w}$
- $(\mathbf{v} + \mathbf{w}) + \mathbf{q} = \mathbf{v} + (\mathbf{w} + \mathbf{q})$ (associativity) $\forall \mathbf{v}, \mathbf{w}, \mathbf{q}$
- $\exists\, \mathbf{0} \in \mathbf{R}^n$ such that $\mathbf{0} + \mathbf{v} = \mathbf{v} + \mathbf{0} = \mathbf{v}$ (null vector) $\forall \mathbf{v}$
- $\forall \mathbf{v}\ \exists \mathbf{v}'$ such that $\mathbf{v} + \mathbf{v}' = \mathbf{0}$
- $1\mathbf{v} = \mathbf{v}, \forall \mathbf{v}$
- $k(m\mathbf{v}) = (km)\mathbf{v}, \forall \mathbf{v} \in \mathbf{R}^n, \forall k \in \mathbf{R}$
- $(k + m)\mathbf{v} = k\mathbf{v} + m\mathbf{v}, \forall \mathbf{v} \in \mathbf{R}^n, \forall k, m \in \mathbf{R}$
- $k(\mathbf{v} + \mathbf{w}) = k\mathbf{v} + k\mathbf{w}, \forall \mathbf{v}, \mathbf{w} \in \mathbf{R}^n, \forall k \in \mathbf{R}$

The above properties are satisfied in the Euclidean space $\mathbf{R}^n$, therefore $\mathbf{R}^n$ is called a *vector space* (or *linear space*) over $\mathbf{R}$. In general, any space with these properties is called a *vector space* (or *linear space*) over $\mathbf{R}$.

The space of vectors and that of scalars may be different from those taken above.

# 10.     PROBABILITY

## 10.1     Borel Algebra

Let $\Omega$ be a set. A set $\mathfrak{I} \subseteq \wp(\Omega)$ of subsets of $\Omega$ is called a $\sigma$–*algebra* (Borel algebra) if the following properties hold ($\forall A, B \in \mathfrak{I}$):

- $\Omega \in \mathfrak{I}$
- $A \cap B \in \mathfrak{I}$ (or $AB \in \mathfrak{I}$)
- $A \cup B \in \mathfrak{I}$ (or $A + B \in \mathfrak{I}$)
- $\Omega \setminus A \in \mathfrak{I}$

## 10.2     Probability

Let $\Omega$ be a set called a *universe* (set of elements called *elementary events*) and $\mathfrak{I} \subseteq \wp(\Omega)$ a $\sigma$–algebra (a set of *events*) over $\Omega$. A *probability measure* (or *probability*) $P$ is defined as follows (Kolmogoroff's axioms):

- $P: \mathfrak{I} \to [0, 1]$
- $P(\Omega) = 1$
- $A \cap B = \varnothing \Rightarrow P(A \cup B) = P(A) + P(B)$  $A, B \in \mathfrak{I}$

## 10.3     Probability and Relative Frequency

Assume that a trial has $n$ possible outcomes all equally likely. If any one of $r$ outcomes produces an event $E$, the *relative frequency* $f_{rel}$ of $E$ is $f_{rel} = r/n$.

It can be shown that the following relationship between the probability $P(E)$ of an event $E$ and its relative frequency $f_{rel}(E)$ exists:

**THEOREM 2.19**

$$\lim_{n \to +\infty} f_{rel}(E) = P(E). \ \blacklozenge$$

In other words, probability is the relative frequency in the long run. A consequence is that, in practice, the probability of an event can be established empirically by performing a large number of trials and equating relative frequency with probability.

EXAMPLE 2.43
What is the probability that in tossing a die the outcome is odd and greater than 2? This event is satisfied by two of the six possible outcomes, namely 3 and 5. Therefore the probability that this event will occur is equal to $2/6 = 1/3$.

## 10.4    Independent Events

Two events $A$ and $B$ are *independent* if

$$P(A \cap B) = P(A) \cdot P(B) \tag{2.69}$$

In words, two trials are said to be independent if the outcome of one trial does not influence the outcome of the other trial.

EXAMPLE 2.44
Tossing a die twice means two independent trials.

## 10.5    Conditional Probability

Let $P(A) > 0$. Then the quantity denoted by $P(B|A)$ and defined as

$$P(B|A) = \frac{P(AB)}{P(A)} \tag{2.70}$$

is called the *conditional probability* of event $B$ relative to event $A$. Alternatively, assume there are two trials, the second dependent on the first. The probability $P(A \text{ and } B)$ that the first trial yields an event $A$ and the second trial yields an event $B$ (which thus is dependent on $A$) is the product of their respective probabilities, where the probability $P(B|A)$ of $B$ is calculated on the premise that $A$ has occurred (the probability of $A$ is $P(A)$):

$$P(AB) = P(A) \cdot P(B|A) \tag{2.71}$$

## 10.6    Bayes' Theorem

Let $A_1, A_2, ..., A_n$ be a mutually disjoint and complete system of events, i.e.

$$A_1 \cup A_2 \cup ... \cup A_n = \Im, A_i \cap A_j = \varnothing, \quad i, j = 1, 2, ..., n, \quad i \neq j$$

and $B$ an arbitrary event. The conditional probability $P(A_i|B)$ is calculated by *Bayes' Theorem* (or *Bayes' Formula*) as follows:

$$P(A_i|B) = \frac{P(B|A_i)P(A_i)}{P(B|A_1)P(A_1) + P(B|A_2)P(A_2) + ... + P(B|A_n)P(A_n)} \qquad (2.72)$$

If $n = 1$ then $P(B) = P(B|A)P(A)$ where $A = A_1$. In this case, Bayes' Theorem becomes

$$P(A|B) = \frac{P(B|A)P(A)}{P(B)} \qquad (2.73)$$

## 10.7    Random Variable

A *random variable* is a real–valued function $\xi: \Im \rightarrow \mathbf{R}$ with the following property:

$$\{A|\xi(A) < x, x \in \mathbf{R}\} \in \Im \qquad (2.74)$$

In words, a random variable is a quantity which assumes values influenced by chance, i.e., they cannot be predetermined.

EXAMPLE 2.45
A box contains 100 objects. Of these, 66 are blue and 34 are red. 20 objects are drawn from the box at random, without being replaced. The number $X$ of blue objects which are drawn is a random variable.

## 10.8    Probability Distribution

Let $\xi$ denote a random variable. The probability $P(\xi < x)$ is called the *probability distribution* $F(x)$ of $\xi$, i.e.,

$$F(x) = P(\xi < x) \tag{2.75}$$

Alternatively, in words, the set of all possible values of $\xi$ and their respective probabilities is called the *probability distribution* of $\xi$.

It can be shown that the following hold:

**THEOREM 2.20**

$F(x) \le F(y)$, if $x < y$;

$\lim\limits_{x \to -\infty} F(x) = 0$ and $\lim\limits_{x \to +\infty} F(x) = 1$ ◆

EXAMPLE 2.46
*Binomial distribution*:

$$P(\xi = k) = {}^{n}C_k p^k (1 - p)^{n-k}, \quad n \in \mathbf{N}, \quad 0 < p < 1 \tag{2.76}$$

i.e. the random variable $\xi$ takes on the values $k = 0, 1, 2, ..., n$ with the given probability $P(\xi = k)$.

## 10.9    Density Function

The *probability–density function* (or *density function*) $f(x)$ of a random variable $\xi$ with probability distribution $F(x)$ is defined as follows:

$$P(\xi < x) = F(x) = \int\limits_{-\infty}^{x} f(t)dt$$

In other words, the graph (curve) of the density function has the following property: the area bounded by the curve, the x–axis, and the vertical lines erected at $X = a$ and $X = b$ is equal to the probability that $a \le \xi \le b$, i.e.,

$$P(a \le \xi \le b) = \int\limits_{a}^{b} f(t)dt$$

EXAMPLE 2.47
a) The random variable $\xi$ has an *exponential distribution* if its density function $f(x)$ is as follows:

$$f(x) = \begin{cases} \mu e^{-\mu x}, x > 0 \\ 0, x \le 0 \end{cases}$$

b) The random variable $\xi$ has a *normal distribution* if its density function $f(x)$ is as follows:

$$f(x) = \frac{1}{\sqrt{2\pi}\sigma} e^{\frac{-(x-m)^2}{2\sigma^2}}$$

c) The random variable $\xi$ has a *uniform distribution* if its density function $f(x)$ is as follows:

$$f(x) = \begin{cases} \frac{1}{b-a}, a \le x \le b \\ 0 \end{cases}$$

## 11.    FUZZY SETS

### 11.1    The Concept of a Fuzzy Set

Let $X$ be a set. A *fuzzy set* $\tilde{A}$ in $X$ is a set of ordered pairs

$$\tilde{A} = \{(x, \mu_{\tilde{A}}(x)) | x \in X\} \tag{2.79}$$

where $\mu: X \to [0, 1]$ is called the *membership function* (or degree of compatibility or truth), meaning a degree to which $x$ belongs to $\tilde{A}$. Elements with a zero degree membership are normally not listed.

EXAMPLE 2.48
Let

$$X = \{1, 10, 15, 18, 24, 40, 66, 80, 100\}$$

be a set denoting possible ages for humans. Then the fuzzy set

Ã = "ages considered as young"

could be

$$\{(10, .5), \ (15, 1), (18, 1), (24, 1), (40, .4)\}$$

If the membership function can only take two values, 0 and 1, the fuzzy set becomes a (*classical* or *crisp*) set: an element either belongs to the set or not.

There are several ways of representing fuzzy sets such as ordered pairs, solely by membership function or by a graph (of membership function).

The membership function need not take values in the interval [0, 1]; any finite interval of non–negative real numbers (or whose supremum is finite) can be considered. If the membership function takes values in the interval [0, 1] the fuzzy set is called a *normalised* fuzzy set. Because any membership function can be normalised it is convenient to consider normalised fuzzy sets.

## 11.2     α cut

Given a fuzzy set Ã in $X$ and a real number $\alpha \in \mathbf{R}$. The (crisp) set

$$A_\alpha = \{x \in X \mid \mu_{\tilde{A}}(x) \geq \alpha\} \tag{2.80}$$

is called an α *level set* or α *cut*. The (crisp) set

$$A_\alpha = \{x \in X \mid \mu_{\tilde{A}}(x) > \alpha\} \tag{2.81}$$

is called a *strong* α *level set* or *strong* α *cut*.

EXAMPLE 2.49
The strong 0.5–cut $A_{0.5}$ of the fuzzy set Ã = "ages considered as young" = $\{(10, .5), \ (15, 1), (18, 1), (24, 1), (40, .4)\}$ is $A_{0.5} = (15, 18, 24)$.

## 11.3     Fuzzy Union

Given two fuzzy sets $Ã_1$ and $Ã_2$ in $X$ with membership functions $\mu_1$ and $\mu_2$, respectively. The membership function $\mu$ of the (fuzzy) *union* $Ã = Ã_1 \cup Ã_2$ is defined as follows:

$$\mu(x) = \max \{\mu_1(x), \mu_2(x)\}, \ \forall x \in X \tag{2.82}$$

## 11.4    Fuzzy Intersection

Given two fuzzy sets $\tilde{A}_1$ and $\tilde{A}_2$ with membership functions $\mu_1$ and $\mu_2$, respectively.

The membership function $\mu$ of the (fuzzy) *intersection* $\tilde{A} = \tilde{A}_1 \cap \tilde{A}_2$ is defined as follows:

$$\mu(x) = \min \{\mu_1(x), \mu_2(x)\}, \forall x \in X \tag{2.83}$$

EXAMPLE 2.50
Let

$$\tilde{A}_1 = \{(10, .5), (15, 1), (18, 1), (24, 1), (40, .4)\}$$

and

$$\tilde{A}_2 = \{(24, .1), (40, .3), (70, .9)\}$$

be two fuzzy sets. Then their fuzzy union is

$$\tilde{A}_1 \cup \tilde{A}_2 = \{(10, .5), (15, 1), (18, 1), (24, 1), (40, .4), (70, .9)\}$$

and their fuzzy intersection is

$$\tilde{A}_1 \cap \tilde{A}_2 = \{(24, .1), (40, .3)\}$$

## 11.5    Fuzzy Complement

The membership function $\mu_{\mathfrak{c}\tilde{A}}(x)$ of the *fuzzy complement* of a normalised fuzzy set $\tilde{A}$ is defined as follows:

$$\mu_{\mathfrak{c}\tilde{A}}(x) = 1 - \mu_{\tilde{A}}(x) \tag{2.84}$$

EXAMPLE 2.51
The fuzzy complement of $\tilde{A}$ (Ex. 2.48) is

$$\{(10, .5), (15, 0), (18, 0), (24, 0), (40, .6)\}$$

## 11.6 Fuzzy Relation

A (binary) *fuzzy relation* $\tilde{A}$ in $X \times Y$ is defined as follows:

$$\tilde{A} = \{((x, y), \mu_{\tilde{A}}(x, y)) \mid (x, y) \in X \times Y\} \qquad (2.85)$$

EXAMPLE 2.52
Let $X, Y \subseteq \mathbf{R}$. Then

$$\tilde{A} = \text{"much larger than"} = \{((100, 0), .8), ((5, 4), .2)\}$$

is a fuzzy relation.

## 11.7 Fuzzy Projections

Let

$$\tilde{A} = \{((x, y), \mu_{\tilde{A}}(x, y)) \mid (x, y) \in X \times Y\}$$

be a fuzzy relation. The *first projection* $\tilde{A}^{(1)}$ of $\tilde{A}$ is defined as follows:

$$\tilde{A}^{(1)} = \{(x, \max_y \mu_{\tilde{A}}(x, y)) \mid (x, y) \in X \times Y\} \qquad (2.86)$$

and the *second projection* $\tilde{A}^{(2)}$ of $\tilde{A}$ is defined as follows:

$$\tilde{A}^{(2)} = \{(y, \max_x \mu_{\tilde{A}}(x, y)) \mid (x, y) \in X \times Y\} \qquad (2.87)$$

EXAMPLE 2.53
Given the following fuzzy relation. The projections $\tilde{A}^{(1)}$ and $\tilde{A}^{(2)}$ are:

|       | $y_1$ | $y_2$ | $\tilde{A}^{(1)}$ (maximum on rows) |
|-------|-------|-------|------------------------------------|
| $x_1$ | .2    | .5    | .5                                 |
| $x_2$ | .6    | 1     | 1                                  |
|       | .6    | 1     | $\tilde{A}^{(2)}$                  |

(maximum on columns)

## 12.      METRIC SPACES

### 12.1      Metric; Metric Space

Given a set $X$, a function $d: X \to \mathbf{R}_+$ and the following properties:

- $d(x, x) = 0,\ \forall x \in X$ (reflexivity)
- $d(x, y) = d(y, x),\ \forall x, y \in X$ (symmetry)
- $d(x, y) + d(y, z) \geq d(x, z),\ \forall x, y, z \in X$ (triangle inequality)
- $d(x, y) = 0 \Rightarrow x = y,\ \forall x, y \in X$ (indiscernibility)

If all properties hold, $d$ is called a *metric* (*distance*) and the pair $(X, d)$ is called a *metric space*.

If all but the last property hold, $d$ is called a *pseudo–metric* and the pair $(X, d)$ is called a *pseudo–metric space*.

EXAMPLE 2.54
Given $d(\mathbf{x}, \mathbf{y})$, $\mathbf{x} = (x_1, x_2, ..., x_n)$, $\mathbf{y} = (y_1, y_2, ..., y_n)$, $n \geq 1$.

a) Euclidean distance ($\mathbf{R}^n$ metric space or Euclidean space):

$$\sqrt{\sum_{i=1}^{n}(x_i - y_i)^2}$$

b) $d(\mathbf{x}, \mathbf{y}) = \max_i |x_i - y_i|$

The concept of a metric is a generalisation of the usual Euclidean distance. It can be shown that:

THEREOM 2.21
Every set can be transformed into a metric space, e.g., with $d(x, y) = 1$ for $x \neq y$ and $d(x, x) = 0$. ◆

### 12.2      Neighbourhood (Vicinity)

Let $X$ be a metric space with metric $d$ and $x \in X$ an arbitrary point. The $\varepsilon$–*ball* (*sphere*) $S(x, \varepsilon)$ of point $x$ is defined as the set of points which lie within a distance $\varepsilon$ from $x$ as follows:

$$S(x, \varepsilon) = \{ y \mid d(x, y) < \varepsilon \} \qquad (2.88)$$

EXAMPLE 2.55
Let $X = \mathbf{R}$ with the Euclidean distance $d(x, y) = |x - y|$. Then $S(1, .5)$ of point $x = 1$ is the open interval $S(1, .5) = (.5, 1.5)$.

A *neighbourhood* (*vicinity*) $V_x \subseteq X$ of point $x$ is defined as a set containing an $\varepsilon$–ball:

$$\exists \varepsilon > 0 \text{ such that } S(x, \varepsilon) \subseteq V_x \qquad (2.89)$$

EXAMPLE 2.56
Let $X = \mathbf{R}$ with the Euclidean distance $d(x, y) = |x - y|$. Then $V_1$ of point $x = 1$ is $V_1 = (0, 2)$ because $S(1, .5) \subseteq V_1$.

## 12.3　Open Set

Given a metric space $X$. A subset $Q \subseteq X$ is called an *open set* if $Q$ is a vicinity for each one of its elements, i.e.:

$$\forall x \in Q \; \exists \, S(x, \varepsilon) \text{ such } S(x, \varepsilon) \subseteq Q \qquad (2.90)$$

$X$ and the empty set $\varnothing$ are open sets. It can be shown that:

**THEOREM 2.22**
The union of an arbitrary number of open sets is open, the intersection of a finite number of open sets is open, and any $S(x, \varepsilon)$ is open, $\forall x, \varepsilon$. ♦

## 12.4　Convergence

Given a metric space $X$ with metric $d$, and a sequence of points $x_i \in X$, $i = 1$, $2, ..., n, ...$ The sequence $(x_i)$ has the limit $x$, notation $x_i \to x$, if

$$x_i \to x \Leftrightarrow (\forall V_x \; \exists m \in \mathbf{N} \text{ such that } x_i \in V_x, \forall i \geq m) \qquad (2.91)$$

In words, the sequence $(x_i)$ tends to limit $x$ if any vicinity $V_x$ of $x$ contains every element of the sequence (apart from, perhaps, a finite number of elements).

## 12.5    Completeness

The natural generalisation of the concept of a Cauchy–sequence to metric spaces is as follows. A sequence $x_n$ is called a *Cauchy–sequence* if

$$\forall \varepsilon > 0 \ \exists n_\varepsilon \in \mathbf{N} \text{ such that } (\forall n, m > n_\varepsilon \Rightarrow d(x_n, x_m) < \varepsilon) \qquad (2.92)$$

A metric space $(X, d)$ is said to be *complete* if every Cauchy–sequence in $X$ has a limit. Any complete metric space is called a *Baire space*.

EXAMPLE 2.57
The Euclidean space $\mathbf{R}^n$ is complete.

It can be shown that:

**THEOREM 2.23**
A (pseudo–)metric space $(X, d)$ is complete if and only if from $S(x_{n+1}, \varepsilon(n+1)) \supseteq S(x_n, \varepsilon(n))$, $n = 1, 2, ...$, where $S(x_n, \varepsilon(n)) = \{ y \mid d(x, y) \leq \varepsilon(n)\}$, and $\varepsilon(n) \to 0$ as $n \to +\infty$, it follows that $\bigcap_n S(x_n, \varepsilon(n)) \neq \varnothing$ ♦

## 12.6    Fixed Point

The point $x \in X$ is called a *fixed point* of a function $f: X \to X$ if

$$f(x) = x \qquad (2.93)$$

It can be shown that the following holds.

**THEOREM 2.24**
Given a complete metric space $(X, d)$, and let $f: X \to X$ be a function. If $\exists \alpha \in \mathbf{R}, 0 < \alpha < 1$, such that $d(f(x), f(y)) \leq \alpha d(x, y), \forall x, y \in X$, then $f$ has a unique fixed point. ♦

## 13.    TOPOLOGY

## 13.1    Topology; Topological Space

Given a set $X$. Let $\Im \subseteq \wp(X)$ be a subset with the following properties:

1. $\varnothing \in \mathfrak{I}$ (the empty set belongs to $\mathfrak{I}$)
2. $X \in \mathfrak{I}$ (the set $X$ itself belongs to $\mathfrak{I}$)
3. $\underset{j \in J}{\bigcup} \tau_j \in \mathfrak{I}, \forall \tau_j \in \mathfrak{I}$

(the union of any family of elements of $\mathfrak{I}$ belongs to $\mathfrak{I}$)

4. $\underset{i=1,\, ,n}{\bigcap} \tau_i \in \mathfrak{I}, \forall \tau_i \in \mathfrak{I}$

(the intersection of any finite number of elements $\mathfrak{I}$ belongs to $\mathfrak{I}$).

Such a $\mathfrak{I}$ is called a *topology* on $X$, and the pair $(X, \mathfrak{I})$ is called a *topological space*. Any member of $\mathfrak{I}$ is referred to as an *open set* of the topological space.

EXAMPLE 2.58
$\mathfrak{I} = \wp(X)$ is a topological space.
$\mathfrak{I} = \{\varnothing, X\}$ is a topological space (*indiscrete topology*).

Let $(X, d)$ be a metric space. The open sets define a topology on $X$. This topology is referred to as the *topology induced by metric*. For example, the Euclidean space $\mathbf{R}^n$ is a topological space with a a topology induced by the Euclidean distance.

## 13.2 Vicinity

Given a topological space $X$ and a point $x \in X$. Any open set of $X$ that contains the point $x$ is referred to as an *open vicinity* of $x$. Any subset of $X$ that contains an open vicinity of $x$ is called a *vicinity* of $x$.

It can be shown that:

**THEOREM 2.25**
A subset of a topological space is open if and only if it is the vicinity of each of its points. ♦

## 13.3 Hausdorff Space

A topological space $X$ is called a *Hausdorff space* if any two distinct points of $X$ have disjoint vicinities, i.e.,

$$(\forall x, y \in X, x \neq y) \Rightarrow (\exists\, V_x, V_y \text{ such that } V_x \cap V_y = \varnothing) \tag{2.94}$$

It can be shown that:

**THEOREM 2.26**
Every (pseudo–)metric space is a Hausdorff space. ♦

## 13.4    Compactness

A topological space $X$ is said to be *compact* if every open cover of $X$ has a
finite subcover, i.e., a finite sub–collection which also covers $X$.

EXAMPLE. 2.59
The space $\mathbf{I}^n = [0, 1]^n$ is a topological space with the standard topology
consisting of all the intervals $[0, b)$, $(a, b)$ and $(a, 1]$, $0 < a, b < 1$. $\mathbf{I}^n$ is
compact relative to the standard topology. The Euclidean space $\mathbf{E}^n$ is not
compact.

# 14. GRAPH THEORY

## 14.1 Basic Concepts

Given a non–empty finite set

$$V = \{v_1, v_2, ..., v_n\} \tag{2.95}$$

of elements called *vertices* (*points*, *nodes*). Let

$$E = \{(v_i, v_j) \mid v_i, v_j \in V\} \tag{2.96}$$

be a set of unordered pairs of distinct vertices called *edges* (directed edges are called *arcs*, undirected ones *lines*).

A graph $G$ is the system of vertices and edges, i.e.,

$$G = (V, E) \tag{2.97}$$

An edge $(v_i, v_j)$ is usually denoted by a letter, e.g., $u_{ij}$.

$G' = (V', E')$ is a *subgraph* of $G = (V, E)$ if $V' \subseteq V$ and $E' \subseteq E$ consisting of the corresponding edges.

A *loop* is an edge connecting a node to itself. A loop, however, is not allowed, by definition, in a graph.

If two verteces are connected by more than one edge, the graph is called a *multigraph*.

A multigraph with loops is referred to as a *pseudograph*.

A graph $G = (V, E)$ is *directed* (*digraph*) if $E$ contains ordered pairs, and *undirected* otherwise. The directed edges are referred to as *arcs*.

A digraph with no symmetric pairs of arcs is called an *oriented* graph.

If edges $u_{ij}$, $i, j = 1, 2, ..., n$, are associated (usually) real values $w_{ij}$, $i, j = 1, 2, ..., n$, the graph is called a *weighted graph* and $w_{ij}$ are called *weights*.

## 14.2 Walk

A *walk* $P$ of a graph $G$ consists of a sequence of vertices and edges beginning and ending with vertices as follows:

$$v_0, u_{01}, v_1, ..., v_i, u_{ij}, v_j, ..., v_{n-1}, u_{n-1,n}, v_n \tag{2.98}$$

Because the edges are evident by context, the walk can be denoted only by its vertices:

$$P = v_0, v_1, ..., v_i, ..., v_n \tag{2.99}$$

A walk $P$ is *closed* if $v_0 = v_n$ and *open* otherwise. It is a *trail* if all the edges are distinct, and a *path* if all the vertices (and thus all the edges) are distinct.

A closed path $P$ with $n > 2$ is called a *cycle*. A graph with no cycles is said to be *acyclic*.

The *length* of a walk is the number of occurrences of edges in it.

The *girth* of a graph $G$ is the length of the shortest cycle (if any) and is denoted by $g(G)$.

## 14.3    Connected Graph

A graph $G$ is *connected* if every pair of vertices are joined by a path.

### 14.3.1    Blocks

A *cut point* of a connected graph $G$ is a vertex whose removal results in disconnecting the graph. The *bridge* is such an edge.

A graph $G$ is said to be *nonseparable* if it is connected and has no cut points. A *block* of a graph is a maximal nonseparable subgraph.

EXAMPLE 2.60
Figure 2.1 shows a cut point $v$ and blocks.

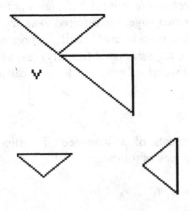

*Figure 2.1* Cut point and block

It can be shown that:

**THEOREM 2.27**
A vertex $v$ of a connected graph $G$ is a cut point if and only if there exist vertices $u$ and $w$ different from $v$ such that $v$ is on every path from $u$ to $w$. ◆

**THEOREM 2.28**
An edge $u$ of a connected graph $G$ is a bridge if and only if it is not on any cycle of $G$. ◆

**THEOREM 2.29**
Every (nontrivial) connected graph has at least two vertices which are not cut points. ◆

## 14.4    Block Graph

The *block graph* of a graph $G$ is denoted by $B(G)$ and defined as follows: every block of $G$ corresponds to a vertex in $B(G)$, and two vertices in $B(G)$ are linked by an edge if the corresponding blocks in $G$ contain a common cutpoint of $G$.

## 14.5    Tree

A *tree* is a connected acyclic graph. It follows that every two points of a tree are joined by a unique path. Figure 2.2 shows a tree.

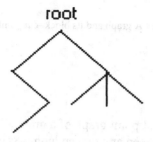

Figure 2.2 A tree

### 14.5.1    Block–Cut Point Tree

Given a connected graph $G$ with blocks $\{b_i\}$ and cut points $\{c_k\}$. The *block–cut point graph* of $G$ is denoted by $bc(G)$ and defined as follows: the vertices of $bc(G)$ correspond to $\{b_i\} \cup \{c_k\}$, and two vertices in $bc(G)$ are linked by an edge if one corresponds to a block $b_i$ and the other to a cut point $c_j$ of $b_i$ (Figure 2.3).

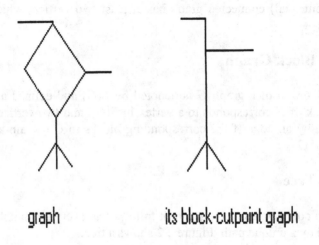

graph                         its block-cutpoint graph

*Figure 2.3* A graph and its block–cut point graph

It can be shown that:

**THEOREM 2.30**

A graph $H$ is the block–cut point graph of a graph $G$ if and only if it is a tree in which the distance between any two endpoints is even. ♦

# 15. MATROID THEORY

## 15.1 Matroid

Given a finite set $E$ and a family

$$C = \{c_i \mid c_i \subseteq E, i = 1, 2, ..., n\} \qquad (2.100)$$

of subsets of $E$ called *circuits*. A *matroid* $M$ is an ordered pair $M = (E, C)$ if the following properties hold:

- no proper subset of a circuit is a circuit;
- $(x \in c_1 \cap c_2) \Rightarrow (c_1 \cup c_2 \setminus \{x\}$ contains a circuit$)$

$E$ is called the *ground set* and the members of $C$ are called the *independent sets*. A subset of $E$ which is not a member of $C$ is called *dependent*.

## 15.2 Cycle Matroid

Given a graph $G = (V, E)$. It can be shown that:

**THEOREM 2.31**
The pair $M = (E, C)$, where $C$ is the set of edge sets of all cycles of $G$, is a matroid. ◆

$M$ is called the *cycle matroid* associated to graph $G$ (Figure 2.4).

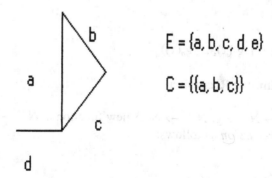

$$E = \{a, b, c, d, e\}$$

$$C = \{\{a, b, c\}\}$$

*Figure 2.4.* Graph and its cycle matroid

# 16.    RECURSION AND COMPLEXITY THEORY

## 16.1    Primitive Recursive Function

The *base functions* are as follows:
a) The *zero* function:

$$0: \mathbf{N} \to \mathbf{N}, \; 0(x) = 0 \tag{2.101}$$

b) The *successor* function:

$$S: \mathbf{N} \to \mathbf{N}, \; S(x) = x + 1 \tag{2.102}$$

c) The *projection* functions:

$$^{n}\pi_{m}: \mathbf{N}^{n} \to \mathbf{N}, \; n \geq m \geq 1 \tag{2.103}$$

$$^{n}\pi_{m}(x_{1}, \ldots, x_{m}, \ldots, x_{n}) = x_{m}$$

## 16.2    Function–forming Operations

### 16.2.1    Composition

The *composition* of $f: \mathbf{N}^{n} \to \mathbf{N}$ and $g_{i}: \mathbf{N}^{m} \to \mathbf{N}$, $i = 1, 2, \ldots, n$, is the following function:

$$h: \mathbf{N}^{n} \to \mathbf{N} \tag{2.104}$$

$$h(x) = f(g_{1}(x), g_{2}(x), \ldots, g_{n}(x))$$

### 16.2.2    Primitive Recursion

Given $f: \mathbf{N}^{n} \to \mathbf{N}$ and $g: \mathbf{N}^{n+2} \to \mathbf{N}$. A new function $h: \mathbf{N}^{m+1} \to \mathbf{N}$ is defined by *primitive recursion* as follows:

$$h(x, 0) = f(x) \tag{2.105}$$

$$h(x, y +1) = g(x, y, h(x, y))$$

## 16.3 Primitive (partial) Recursive Function

The *primitive recursive functions* are the set containing the zero, successor and projection functions, and closed under composition and primitive recursion.

EXAMPLE 2.61
Addition: $f(x, 0) = x$, $f(x, y + 1) = f(x, y) + 1$
Multiplication: $f(x, 0) = 0$, $f(x, y + 1) = f(x, y) + x$

## 16.4 Recursive Function

Let $f: \mathbf{N}^{n+1} \to \mathbf{N}$ be a primitive recursive function. A new function $g: \mathbf{N}^n \to \mathbf{N}$ is defined by *minimising f* as follows:

$$g(x) = \min y \quad \text{such that} \quad f(x, y) = 0 \tag{2.106}$$

The *recursive functions* are the set containing the zero, successor and projection functions, and closed under composition, primitive recursion and minimisation.

## 16.5 Turing Machine

The *Turing machine* is a mathematical abstraction of computation (procedure, program) in general.
Given a set

$$\Sigma = \{s_0, s_1, ..., s_p\} \tag{2.107}$$

called *alphabet*, a set of *states*

$$S = \{q_0, q_1, ..., q_m, q_Y, q_N\}, \quad q_0 = \text{start state}, \quad q_Y, q_N = \text{halt states} \tag{2.108}$$

and a set $M$ (*program*) of *instructions*

$$I = \{q_i s_j s_k X q_j\} \tag{2.109}$$

where $X$ can be $L$ meaning 'to the left' or $R$ meaning 'to the right' or $N$ meaning 'remain in place'. The machine has a linear tape which is potentially infinite in both directions, divided into boxes or cells, and read by a reading head which scans one cell at a time. Symbols are written on the tape, one symbol in one cell.

The machine operates in (*time*) *steps* according to prescribed rules (procedure or program) as follows: the symbol $q_j$ indicates the state the machine is in, $s_j$ is the tape symbol the head is reading, $s_k$ is the symbol that will replace $s_j$, if $X = L$ the head will move one cell to the left, if $X = R$ the head will move one cell to the right, and if $X = N$ the head will remain on the cell where it is.

## 16.6    Turing computability

The Turing machine can be used to define a concept of computability. The machine computes function as follows. Given a function $f: \mathbf{N}^n \rightarrow \mathbf{N}$. We say that $f$ is *Turing computable* if the machine halts with a representation of $f(x)$ on the tape when it starts with a (appropriate) representation of $x$ on the tape.

It is widely accepted today that any function that can be computed by an algorithm (procedure) can also be computed by an appropriate Turing machine (*Church–Turing thesis*).

## 16.7    Recursion and Computability

An important link between recursion and Turing machines is that it can be shown that

**THEOREM 2.32**
$f$ is recursive $\Leftrightarrow f$ is Turing computable. ♦

Because of the above equivalence and with reference to the Church–Turing thesis, it can be hypothesised that the concepts of

- algorithm,
- Turing machine,
- recursive function

are equivalent.

## 16.8    Recursive and Recursively Enumerable Set

A set $X \subseteq Y$ is said to be *recursively enumerable* (or Diophantine), denoted r.e., if there is an algorithm (Turing machine, recursive function, see above) which outputs 'yes' (or 1, say) when presented with an element $x \in Y$ if and only if $x \in X$. If $x \notin X$ the algorithm is undecidable (will never terminate).

A set $X \subseteq Y$ is said to be *recursive* (decidable, computable), if there is an algorithm (Turing machine, recursive function, see above) which outputs 'yes' (or 1, say) when presented with an element $x \in Y$ if $x \in X$ and 'no' (or 0, say) if $x \in Y \setminus X$.

It can be shown that:

**THEOREM 2.33**
For any $A \subseteq \mathbf{N}$ the following equivalence holds:
$A$ is r.e. $\Leftrightarrow$ $A$ is the domain of definition of a primitive recursive function. ◆

In other words, a set whose elements are given (generated) by a primitive recursive function is r.e.. Of course, the set $A$ can be infinite and its elements cannot be effectively generated but we can be sure that any member of it can potentially be generated.

## 16.9    Fixed Point

### 16.9.1    Index

Let $P$ denote a program (or equivalently an algorithm or a recursive function; see above). Formally, $P$ can be conceived as being a string of symbols

$$s_1, s_2, ..., s_i, ..., s_n \tag{2.110}$$

where the symbols $s_i$ are drawn from some base set of *primitive symbols*, e.g.,

$$0 f \sim \vee \Pi \, ( \, ) \, x_1 \, y_3 \, ...$$

Each primitive symbol can be assigned a natural number as follows:

| 0 | $f$ | $\sim$ | $\vee$ | $\Pi$ | ( | ) | $x_1$ | $y_n$ | ... |
|---|-----|--------|--------|-------|---|---|-------|-------|-----|
| 1 | 3 | 5 | 7 | 11 | 13 | 17 | $19^1$ | $23^n$ | ... |

Let $g_i$ be the number assigned to the primitive symbol $s_i$, $i = 1, 2, ..., n$, in the above table. Then a program

$$P = s_1, s_2, ..., s_i, ..., s_n \tag{2.111}$$

is assigned a unique natural number $G$, called the *Gödel number*, as follows:

$$2^{g_1} \cdot 3^{g_2} \cdots p_n^{g_n}$$

where $p_i$ is the $i$th prime number ($p_1 = 2$, $p_2 = 3$, ..., $p_n$).

EXAMPLE 2.62
The string

$$x_1 \vee \sim f$$

is assigned the following Gödel number

$$G = 2^{19} 3^7 5^5 7^3$$

The Gödel numbering provides a method that associates a unique natural number to a program. Moreover, given any natural number $m$, the method allows one to decide whether there is a program whose Gödel number is $m$, and also to identify that program.

The Gödel number $G$ associated with a program $P$ can be viewed as a *coding* of $P$. This number is called an *index* of $P$ and denoted by $e$. Equivalently, we can speak about the index $e$ of an algorithm or recursive function (they being equivalent concepts; see above).

Note
Obviouly, the Gödel number is not the only way to code $P$.

### 16.9.2    Fixed Point

It can be shown that the following hold:

THEOREM 2.34 (Rogers' Fixed Point Theorem)
Let $f$ be a computable (primitive recursive) function. Then there exists an index $e$ such that $\{e\}(x) = \{f(e)\}(x)$, for any input $x$. ♦

In other words, $e$ is a fixed point of $f$. Rogers' Fixed Point Theorem gives the existence of a fixed point.

**THEOREM 2.35** (First Recursion Theorem)

Let $\Phi: F_m \to F_m$ be a recursive function where $F_m$ denotes the primitive recursive functions $N^m \to N$. Then $\Phi$ has a least fixed point and this is computable. ◆

## 16.10   The P Class

Given a sequence $s$ of symbols from the input alphabet $\Sigma$, notation: $s \in \Sigma^*$. A program $M$ accepts s if and only if $M$ halts in the state $q_Y$ when applied to the input $s$.

A language $L_M$ recognised by a program $M$ is defined as follows:

$L_M = \{s \in \Sigma^* | M$ accepts s$\}$ is the language *recognised* by program $M$.

A computational complexity $T_M(n)$ of a program (algorithm) $M$ is defined as follows. For a program $M$ that halts for all $s \in \Sigma^*$, its *time complexity* $T_M(n)$ is

$T_M(n) = \max \{m | \exists s \in \Sigma^*, \text{length}(s) = n, s$ is accepted in $m$ steps$\}$.

The intuitive notion of feasible or tractable programs is expressed in the concepts of **P**–Class which can now be defined as follows:

**P**–Class $= \{L | \exists M$ such that $(L = L_M) \land (\exists p(n)$ polynomial: $T_M(n) \leq p(n)$, $\forall n \in N\}$.

It can be shown that:

**THEOREM 2.36**

Common algorithms used in computational practice (e.g. Quick Sorting, Binary Searching, etc.) belong to the **P**–Class. ◆

## 16.11   The NP Class

Intuitively, the **NP** Class contains those — usually optimisation — problems for which algorithms are intractable; for example, there time complexity grows exponentially with input.

EXAMPLE 2.63 (THE TRAVELLING SALESMAN PROBLEM, TSP)

Given a set $\{c_1, c_2, ..., c_i, ..., c_n\}$ of cities and a distance matrix $D = (d_{ij})_{n,n}$ where $d_{ij}$ denotes a *distance* between the cities $c_i$ and $c_j$ ($d_{ii} = 0$, $d_{ij} = d_{ji}$). To find the *shortest tour* of the cities, i.e.,

$$\min c(\text{TSP}) = \min \left( \sum_{i=1}^{m} d_{\pi(i)\pi(i+1)} + d_{\pi(n)\pi(1)} \right), m = n - 1$$

where $\pi$ denotes a tour (an ordering) of cities.

It can be shown that any optimisation problem has an equivalent expression in the form of a *decision problem* whose solution is either 'yes' or 'no'.

EXAMPLE 2.64
The TSP above can be turned into a decision problem as follows: given a TSP as above as well as a positive integer $B$. Is there a tour $\pi$ such that $c(\text{TSP}) \leq B$?

It can be shown that any instance of a decision problem can be *encoded* using an appropriate set $\Sigma$ of symbols (alphabet). Thus the original optimisation problem, from a formal point of view, is turned into a sequence $s$ of symbols which may be presented to a Turing machine for recognition (acceptance).

This idea helps define the **NP** Class. A special Turing–machine, called a *Nondeterministic Turing machine* (*NDTM*) is constructed. The NDTM is obtained by extending the Turing–machine with a *guessing module* (*GM*) which is a mathematical counterpart of intuitively trying to guess a solution of the problem at hand. After the *GM* has produced, at random, a *candidate* solution, the machine operates as a classical Turing machine in order to *check* the candidate solution, i.e., it is trying to accept the problem instance. The problem instance is accepted if the machine halts in state $q_Y$. The language $L_M$ recognised by an *NDTM* and $m$ are defined as above. The time complexity is defined ad follows:

$$T_M(n) = \max \{1, m\}.$$

The **NP** Class is defined as follows. Given an *NDTM*. The **NP** Class is as follows:

$$\textbf{NP Class} = \{L | \exists M \text{ such that } (L = L_M) \wedge (\exists p(n) \text{ polynomial}: T_M(n) \leq p(n),$$
$$\forall n \in \mathbf{N}\}.$$

It can be shown that the following hold:

THEOREM 2.37
Given any problem in the **NP** Class. Then there exists an exponential time complexity program on a deterministic Turing machine which can solve it. ♦

Problems in the **NP** Class which are equivalent to each other are said to be **NP** *Complete*.

## 17.    ARTIFICIAL NEURAL NETWORK

### 17.1    Artificial Neuron

Let

$$N = \{v_1, v_2, ..., v_i, ..., v_n\} \qquad (2.112)$$

be a set of elements called *artificial neurons* or *neurons*, for short. Any neuron $v_i$

- has associated a real–valued time–dependent quantity called *state* and denoted by $u_i(t)$ where $t$ denotes *time* (or *step*);
- can have an *input* which is denoted by $I_i$;
- outputs a value $f_i(u_i)$ depending on its state $u_i$ and according to a *transfer function* $f_i$ (Figure 2.5).

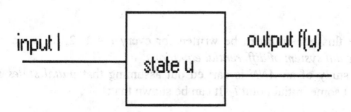

*Figure 2.5* An artificial neuron

### 17.2    Artificial Neural Network

The artificial neurons can be connected to each other, thus forming an *Artificial Neural Network* (*ANN*). Given two interconnected neurons $v_i$ and

$v_j$ in an *ANN*, the output $f_j(u_j)$ of $v_j$ is transferred to $v_i$ via the connection between them which can alter $f_j(u_j)$ by a factor $T_{ij}$ called *weight*. Thus, $T_{ij} \cdot f_j(u_j)$ reaches $u_i$ (Figure 2.6).

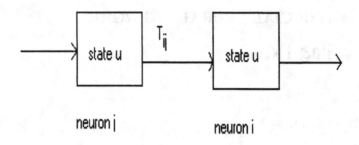

*Figure 2.6* Interconnected neurons

## 17.3    The Fundamental State Equation

The following differential equation can be derived for the state $u_i$ of neuron $v_i$:

$$\frac{du_i(t)}{dt} = -u_i(t) + \sum_{j=1}^{n} T_{ij}f_j(u_j(t)) + I_i(t) \qquad (2.113)$$

Because this equation can be written for every $i = 1, 2, ..., n$ we have a *fundamental system of differential equations*.

The study of an *ANN* is carried out assuming that *initial states* $u_0$ are known at some initial point $t_0$. It can be shown that:

**THEOREM 2.38**
In a small enough vicinity $|u - u_0|$ of $u_0$ and $|t - t_0|$ of $t_0$, the fundamental system has a unique solution. ♦

## 17.4    Operation

The fundamental equations give the state of every neuron at time $t$. By letting time $t$ to evolve, a sequence $u$ of states results. This is referred to as the *operation* of *ANN*.

## 17.5    Energy Function

An *energy function* $E(t)$ associated to an *ANN* is a quantity which decreases with time when the networks operates. It is defined as follows:

$$E(t) = -0.5 \sum_{i,j=1}^{n} T_{ij} f_i(u_i) f_j(u_j) - \sum_{i=1}^{n} I_i f_i(u_i) + \sum_{i=1}^{n} S_i \qquad (2.114)$$

where $S_i$ is

$$\int_{0}^{f_i(u_i)} f_i^{-1}(x)dx$$

It can be shown that:

**THEOREM 2.39**

If $T_{ij} = T_{ji}$ (symmetry) $\forall i, j$ and $T_{ii} = 0 \ \forall i$, and the functions $f_j$, $\forall j$, are monotonically increasing, then $dE/dt < 0$, showing that energy decreases with time. ◆

It can also be seen that $dE/dt = 0 \Rightarrow du/dt = 0$ which means that stationary points for the energy are stationary points for the entire *ANN*.

## 17.6    Equilibrium and Stability

Normally, an *ANN* evolves in time towards a state that does not change anymore. This is called an *equilibrium* and defined as follows:

$$\frac{du_i}{dt} = 0, \quad i = 1, 2, ..., n \qquad (2.115)$$

It can be shown that, under certain conditions on the transfer functions and weights, the equilibrium is *asymptotically stable*, i.e., tends to 0 exponentially.

## 17.7    The Winner Takes All Strategy

The principle of the *winner takes all strategy* (*WTA*) is as follows: only the neuron with the highest state will have output above zero, all the others are 'suppressed'. In other words, *WTA* means selecting the neuron which has the maximum state and deactivate all the others:

$$(u_i = 1 \text{ if } u_i = \max_j u_j) \wedge (u_k = 0 \text{ if } u_k \neq \max_j u_j) \qquad (2.116)$$

## 17.8    Learning

Artificial neurons can be grouped together to form structures called *layers*. A collection of layers forms a *processing structure for solving a problem*. Usually, there are the following types of layers (Figure 2.7):

- *input layer*: it accepts the problem input data which is called an *input pattern*; there is usually one input layer,
- *output layer*: it delivers a solution (which is called an *output pattern*) of the problem to be solved, there is usually one output layer,
- *hidden layer*(s): there may be more such layers; they act as intermediary processing ensembles.

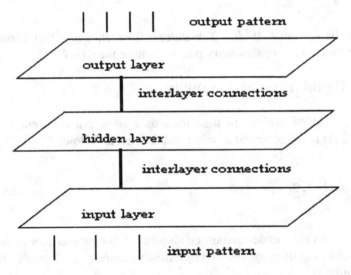

*Figure 2.7* Typical ANN architecture.

Artificial neurons within the same layer usually have the same transfer function, and obey the same *learning rule* which can be defined as an algorithm according to which weights change. Learning rules are based on *learning methods* such as:

- *supervised learning*: both the input pattern and the solution are known; the actual output pattern, produced during the operation of the network, is compared with the solution, and weights are changed accordingly;
- *unsupervised learning*: the network develops its own classification rules;
- *reinforcement learning*: the net is rewarded if the actual output pattern is accepted.

Typically, learning rules are derived from the *Hebb's Rule* which is as follows: if two neurons are simultaneously active, the weight between them is increased by the quantity:

$$k \cdot f_i \cdot f_j \qquad (2.117)$$

Examples of learning rules derived from Hebb's Rule are the *Delta Rule*, the *Kohonen Rule*, etc..

Globally, in a layered *ANN* architecture there is an activation flow between layers:

- *feed forward propagation*: activation flows from the input layer towards the output layer;
- *feed backward propagation*: activation flows from the output layer towards the input layer;
- *interactive propagation*: both forward and backward;
- *equilibrium*: the network relaxes.

When applying *ANN* techniques to *IR*, weights are usually related to, e.g., word frequencies, and special learning rules are derived from Hebb's Rule.

## BIBLIOGRAPHY

Aleksander, I. and Moron, H. (1990). *Neural Computing*. Chapman and Hall.

Aleksander, I. and Moron, H. (1995). *An Introduction to Neural Computing*. Thompson International, London.

Anderson, J.A. and Rosenfeld, E. (1988, eds.). *Neurocomputing Foundations of Research*. The MIT Press.

Arbib, M. A. (1995, ed.). *The Handbook of Brain Theory and Neural Networks*. The MIT Press.

Bibel, W. and Jorrand, Ph. (1988). *Fundamentals of Artificial Intelligence – An Advanced Course*. Lecture Notes in Computer Science, No. 232, Springer–Verlag.

Császár, A. (1978). *General Topology*. Disquisitiones Mathematicae Hungaricae, Budapest, Akadémiai Kiadó.

DeWilde, Ph. (1997). *Neural Network Models*. Springer Verlag London.

Dubois, D. and Prade, H. (1980). *Fuzzy Sets and Systems: Theory and Applications*.

Dugundji, J. (1966). *Topology*. Boston, Allyn and Bacon.

Dumitriu, A. (1977). *History of Logic*. Abacus Press, Kent.

Enderton, H. B. (1972). *A Mathematical Introduction to Logic*. Academic Press, New York.

Garey, M. R. and Johnson, D. S. (1979). *Computers and Intractability*. Bell Telephone Laboratories.

Grossberg, S. (1980). How does a brain build a cognitive code? *Psychological review*, 87, 1–51

Halmos, P. R. (1968). *Naive Mengenlehre*. Vandenhoek & Ruprecht in Goettingen.

Hamilton, A. G. (1978). *Logic for Mathematicians*. Cambridge University Press.

Harary, F. (1972). *Graph Theory*. Addison–Wesley.

Harary, F. (1972). *Graph Theory*. Addison–Wesley Publishing Company.

Hebb, D. (1949). *The Organization of Behavior*. Wiley, New York.

Hopfield, J. J. (1982). Neural networks and physical systems with emergent collective computational abilities. *Proceedings of the National Academy of Sciences*, 79, 2554–2558

Hughes, G. and Cresswell, M. (1968). *An introduction to modal logic*. Methuen, London.

Kelley, J. L. (1975). *General topology. Graduate texts in mathematics, 27*. New York, Springer Verlag.

Klir, G. J. & Yuan, B. (1995). *Fuzzy Sets and Fuzzy Logic*. Prentice Hall PTR.

Kneale, W. and Kneale, M. (1962). *The Development of Logic*. Oxford University Press.

Kohle, M. (1990). *Neuronale Netze*. Springer Verlag.

Kohonen, T. (1972). Correlation matrix memories. *IEEE Transactions on Computers*, 21, 353–359

Kohonen, T. (1988). An introduction to neural computing. *Neural Networks*, 1(1), 3–16.

Krivine, J. L. (1971). *Introduction to axiomatic set theory*. D. Reidel Publishing, Co.

Kurtz, M. (1991). *Handbook of Applied Mathematics for Engineers and Scientists*. McGraw–Hill, Inc., New York.

Lewis, H. R. and Papadimitrou, C. H. (1998). *Elements of the Theory of Computation*. Prentice–Hall.

Makowsky, J. A. (1992). Model Theory and Computer Science: An Appetizer. In: Abramsky, S., Gabbay, D. M. and Maibaum, T. S. E. (1992, eds.). *Handbook of Logic in Computer Science Vol. 1 Backgrounf and Mathematical Structures*. Clarenden Press, Oxford.

McCulloch, W. S. and Pitts, W. (1943). A logical calculus of the ideas immanent in nervous activity. *Bulletin of Mathematical Biophysics*, 5, 115–133

Mendelson, E. (1964). *An introduction to mathematical logic*. Van Nostrand, Reinhold, Princeton, New Jersey.

Minsky, M. and Papert, S. (1969). *Perceptrons*. The MIT Press.

Minty, G. (1966). On the axiomatic foundations of the theories of directed linear graphs, electrical networks and network–programming. *J. Math. Mech.*, 15, 485–520.

Oxley, J. G. (1992). *Matroid Theory*. Oxford University Press, Oxford New York Tokio.

Philips, I. C. C. (1992). Recursion Theory. In: Abramsky, S., Gabbay, D. M. and Maibaum, T. S. E. (eds.) *Handbook of Logic in Computer Science Vol 1 Background: Mathematical Structures*. Oxford Science Publications, Oxford Press – Oxford, 1992, 79–187.

Rogers, H. (1967). *Theory of Recursive Functions and Effective Computability*. McGraw–Hill, New York.

Rosenblatt, F. (1958). The perceptron: a probabilistic model for information storage and organization in the brain. *Psychological Review*, 65, 386–408

Rosser, J. B. and Turquette, A. (1958). *Many–valued logics*. Amsterdam, North Holland.

Rumelhart, D. E., Hinton, G. E. and Williams, R. J. (1986). Learning representations by back–propagating errors. *Nature*, 323, 533–536

Ryan, M. and Sadler, M. (1992). Valuation Systems and Consequence Relations. In: Abramsky, S., Gabbay, D. M. and Maibaum, T. S. E. (1992, eds.). *Handbook of Logic in Computer Science Vol. 1 Background and Mathematical Structures*. Clarenden Press, Oxford.

Schubert, H. (1984). *Topologie – Eine Einführung*. Stuttgart, Teubner Verlag.

Smyth, M. B. (1992). Topology. In: Abramsky, S., Gabbay, D. M. and Maibaum, T. S. E. (1992, eds.). *Handbook of Logic in Computer Science Vol. 1 Backgrounf and Mathematical Structures*. Clarenden Press, Oxford.

Turner, R. (1984). *Logics for artificial intelligence*. Ellis Harwood, Ltd., Chichester, England.

Turner, R. (1984). *Logics for artificial intelligence*. Ellis Harwood, Ltd., Chichester, England.

Tutte, W. T. (1965). Lectures on matroids. *J. Res. Nat. Bur. Stand.*, 69, 1–47.

Willard, S. (1970). *General Topology*. Reading, MA: Addison–Wesley.

Zadeh, L. A. (1965). Fuzzy sets. *Information Control*, 8, 338–353.

Zimmermann, H.-J. (1996). *Fuzzy Set Theory and Its Applications*. Kluwer Academic Publishers, Boston/Dordrecht/London.

# Chapter 3

# INFORMATION RETRIEVAL MODELS

*Information Retrieval* (*IR*) is concerned with the organisation, storage, retrieval, and evaluation of information likely to be relevant to a user's information need.

Information is looked for in objects, traditionally called *documents*, and these may be text, image, sound or multimedia objects. The collection of documents is stored in databases on computer disks.

The user's *information need* is formulated in a *request*, e.g., natural language — for example, "Articles on drogs misuse after 1960.". The request is to be re–formulated in the form of a *query*, i.e., a form that is 'understood' by the computer (as required by the syntax of a query language).

*Retrieval* means looking for answers to queries in the collection of objects, and is in practice given by an algorithm based on mathematical methods and formulae.

In a retrieval algorithm computationally appropriate *representations* of documents and queries are used rather than the original documents and queries themselves.

Depending on how the documents, query and retrieval are modelled, one can distinguish different formal *models* (classes, types) of *IR*.

This chapter describes in a mathematical fashion the basic model types of *IR* developed so far. They are based on ideas and specific techniques from different scientific fields such as:

- Mathematics;
- Logics;
- Information Science;
- Computer Science;

–   Artificial Intelligence (*AI*);
–   Quantum Mechanics (*QM*).

The models based on mathematical techniques were the first models of *IR*, because, on the one hand, the mathematical knowledge used was well known and well understood, and on the other hand, because the models so created were (relatively) easily and immediately implementable. Their mathematical roots can be traced back to the 1950s. In these model types retrieval of information is based on mathematically measuring a 'distance' (similarity, matching, likeness) between — the representations of — a user's query and objects to be searched. Different specific cases of these models are used in virtually all commercial *IR* systems today. This is why the mathematical models of *IR* are considered to be *classical models* of *IR*.

The models of *IR* based on special techniques from logic and information science constitute a relatively newer category. Their roots can be traced back to the 1980s. In these model types retrieval of information is based on whether there is some inference process or flow of information between a user's query and objects to be searched.

The models of *IR* applying different *AI* and *AI*–related methods and procedures (e.g., Artificial Neural Netwroks (*ANN*), Knowledge Bases (*KB*), Natural Language Processing (*NLP*), Genetic Algorithms (GA), etc.) will be referred to as *alternative* models. They enhance the classical models of *IR*. It oftens happens that mixed techniques are applied. In these model types retrieval is interpreted in the language of the particular *AI* field used. Their roots can be traced back to the 1980s.

The last category of models is based on ideas and principles from Quantum Mechanics (*QM*). Their roots go back to the early 1990s. In these model types retrieval of information is a result of an effective and real interaction between a user's query and objects to be searched.

Thus the different *IR* models can also be categorised relative to the classical models, as follows:

•   Classical models of *IR*;
•   Non–classical models of *IR*;
•   Alternative models of *IR*.

In this chapter the above classification will be followed. A basic and representative model will be described in each case using a mathematical formalism. Examples will also be given. Each particular model of *IR* can be separately read, understood, learned, and, if one wishes, implemented (i.e. programmed on computer), although technical and implementation details are not discussed, as these would go beyond the scope of this book. (For algorithmic aspects, see, e.g., Baeza–Yates and Ribeiro–Neto, 1999).

The basic theoretical model types of *IR* reflect, on the one hand, the complexity and interdisciplinary nature of *IR* in general (see, e.g., [Florian and Buckland, 1990; Karamuftuoglu, 1998] and of *IR* modelling in particular (whose manifold nature is also well shown in recent years' few books: [Frants, Shapiro and Voikunskii, 1997; Korfhage, 1997; Kowalski, 1997; Baeza–Yates and Ribeiro–Neto, 1999]), and, on the other hand the modelling difficulty of the most critical modelling parameter which is relevance (although it is mainly a subjective category, its central role in *IR* challenges — inherently imperfect — mathematical modelling for it; see e.g. [Rocchio, 1971; Gordon, 1989, 1990; Froelich, 1991, 1994; JASIS 1994; Buckland and Gey, 1994; Su, 1994; Allan, 1996; Belkin et. al., 1996; Belkin and Koenemann, 1996; Ellis, 1996; Mizzaro, 1997; Shaw, Burgiu and Howell, 1997; Heine, 1990, 1999].

# 1. CLASSICAL MODELS OF INFORMATION RETRIEVAL

The classical *IR* models are as follows:

- Boolean *IR* (*BIR*);
- Vector (or vector space) *IR* (*VIR*);
- Probabilistic *IR* (*PIR*).

They are called so because they were the first models of *IR*, and virtually every commercial *IR* system today is based upon them (and especially upon the Boolean model of *IR*). The classical *IR* models are based on well–elaborated, precise and implementable mathematical methods. This part describes a basic mathematical model for each of these classical models, along with simple examples. Complex examples and procedures are also given in Appendices 1 and 2 (in *MathCAD 8 Professional Plus*).

## 1.1 Boolean Model

The *Boolean model* of *IR* (*BIR*) is a classical *IR* model and, at the same time, the first and most widely adopted one. It is used by virtually all commercial *IR* systems today.

The *BIR* is based on Boolean Logic and classical Set Theory in that both the documents to be searched and the user's query are conceived as sets of terms. Retrieval is based on whether or not the documents contain the query terms.

Given a finite set

$$T = \{t_1, t_2, ..., t_j, ..., t_m\}  \tag{3.1}$$

of elements called *index terms* (e.g. words or expressions — which may be roots of words — describing or characterising documents such as keywords given for a journal article), a finite set

$$D = \{D_1, ..., D_i, ..., D_n\}, D_i \in \wp(T)  \tag{3.2}$$

of elements called *documents*.

Traditionally, a (real) document can be a journal article (or its abstract or title), or a newspaper article, etc.. These documents are formally conceived, for retrieval purposes, in the *BIR*, as being represented by a set of keywords or terms. In practice the orginal document and its representation are two different entities. In principle, however, from a purely formal mathematical point of view, it is not a logical restriction to refer to document representations as documents, for two reasons: i) there is a correspondence between the original document and its representation, and ii) for retrieval purposes the document representation is used rather than the original (although the formalism would be valid).

Given a Boolean expression $Q$ — in a normal form — called a *query* as follows:

$$Q = \bigwedge_{k \in K} ( \bigvee_{j \in J} \theta_j), \ \theta_j \in \{t_j, \neg t_j\}  \tag{3.3}$$

Equivalently, $Q$ can be given in a disjunctive normal form, too.

An operation called *retrieval*, consisting of two steps, is defined as follows:

1. The sets $S_j$ of documents are obtained that contain or not term $t_j$:

$$S_j = \{D_i | \theta_j \in D_i\}  \tag{3.4}$$

where $\neg t_j \in D_i$ should be read $t_j \notin D_i$.

2. Those documents are retrieved in response to $Q$ which are the result of the corresponding sets operations, i.e., the answer to $Q$ is as follows:

$$\bigcap_{k \in K} (\bigcup_{j \in J} S_j) \qquad (3.5)$$

EXAMPLE 3.1

Let the set of original (real) documents be, for example

$$O = \{O_1, O_2, O_3\},$$

where

$O_1$ = *Bayes' Principle: The principle that, in estimating a parameter, one should initially assume that each possible value has equal probability (a uniform prior distribution).*

$O_2$ = *Bayesian Decision Theory: A mathematical theory of decision making which presumes utility and probability functions, and according to which the act to be chosen is the Bayes act, i.e., the one with highest Subjective Expected Utility. If one had unlimited time and calculating power with which to make every decision, this procedure would be the best way to make any decision.*

$O_3$ = *Bayesian Epistemology: A philosophical theory which holds that the epistemic status of a proposition (i.e., how well proven or well established it is) is best measured by a probability and that the proper way to revise this probability is given by Bayesian conditionalisation or similar procedures. A Bayesian epistemologist would use probability to define and explore the relationship between concepts such as epistemic status, support, or explanatory power.*

Let the set $T$ of terms be:

$T = \{t_1 = $ Bayes' Principle, $t_2 = $ probability, $t_3 = $ decision making, $t_4 = $ Bayesian Epistemology$\}$

Then the set $D$ of documents is as follows:

$$D = \{D_1, D_2, D_3\}$$

where

$D_1$ = {Bayes' Principle, probability}

$D_2$ = {probability, decision making}

$D_3$ = {probability, Bayesian Epistemology}

Let the query $Q$ be:

$Q$ = probability $\wedge$ decision making

1. Firstly, the following sets $S_1$ and $S_2$ of documents $D_i$ are obtained (retrieved):

$S_1 = \{D_i| \text{probability} \in D_i\} = \{D_1, D_2, D_3\}$

$S_2 = \{D_i| \text{decision making} \in D_i\} = \{D_2\}$

2. Finally, the following documents $D_i$ are retrieved in response to $Q$:

$$\{D_i|D_i \in S_1 \cap S_2\} = \{D_1, D_2, D_3\} \cap \{D_2\} = \{D_2\} \qquad (3.6)$$

This means that the original document $O_2$ (corresponding to $D_2$) is the answer to $Q$.

Obviously, if there are more than one document with the same representation, every such document is retrieved. Such documents are, in the *BIR*, indistinguishable (or, in other words, equivalent).

From a purely formal mathematical point of view, the *BIR* is straightforward. From a practical point of view, however, several further problems should be solved which relate to e.g. algorithms and data structures, such as, for example, the choice of terms (manual or automatic selection or both), stemming, hash tables, inverted file structure, and so on. In this respect the interested reader is referred to, e.g., [Hays, 1966; Halpern and Shaw, 1969; van Rijsbergen, 1979; Ullman, 1980; Weiss, 1995; Baeza–Yates and Ribeiro–Neto, 1999].

Mathematical properties of *BIR* and its relation to other *IR* models will'be dealt with in Chapter 4.

## 1.2    Vector  Space Model

The *Vector* (or *vector space*) *Model* of *IR* (*VIR*) is a classical model of *IR*. It is — traditionally — called so because both the documents and queries are conceived, for retrieval purposes, as strings of numbers as though they were (mathematical) vectors. Retrieval is based on whether the 'query vector' and the 'document vector' are 'close enough'.

Given a finite set *D* of elements called *documents*:

$$D = \{D_1, ..., D_i, ..., D_n\} \qquad (3.7)$$

and a finite set *T* of elements called *index terms*:

$$T = \{t_1, ..., t_j, ..., t_m\} \qquad (3.8)$$

Any document $D_i$ is assigned a vector $\mathbf{v}_i$ of finite real numbers, called *weights*, of length *m* as follows:

$$\mathbf{v}_i = (w_{1i}, ..., w_{ji}, ..., w_{mi}) \qquad (3.9)$$

where $0 \leq w_{ji} \leq 1$ (i.e., $w_{ji}$ is normalised, e.g., division by the largest).  The weight $w_{ji}$ is interpreted as an extent to which the term $t_j$ 'characterises' document $D_i$.

The choice of terms and weights is a difficult theoretical (e.g., linguistic, semantic) and practical problem, and several techniques can be used to cope with it. A basic technique is described in what follows.

The *VIR* model assumes that the most obvious place where appropriate content identifiers might be found are the documents themselves. In, e.g., [Luhn, 1959; Hays, 1966] it is assumed that frequencies of words in documents can give meaningful indication of their content, hence they can be taken as *identifiers*.

When computing word frequencies and arranging (ranking) the words in decreasing order of their frequency (or alternatively, number of occurrence), then the frequency of a given word multiplied by its rank order is approximately equal to the frequency of another word multiplied by its rank. This is known as the *rank—frequency law* of Zipf (or Zipf's law; see, e.g., Zipf, 1949), which can be written as follows:

$$frequency \times rank \approx constant \qquad (3.10)$$

Taking into account the above it is possible to elaborate different methods to compute word significance factors (or weights) related to documents. In what follows, a simple automatic method is presented which consists of the following steps:

- Exclude *stop words* from documents, i.e., those words that are unlikely to bear any significance. For example, *a, about, and*, etc. Obviously a given application field can have its own list of stopwords not valid in a different area. For example, see the stoplists of standard test collections, e.g., SMART, TIME, ADI, Cranfield.
- Apply *stemming* to the remaining words, i.e. reduce or transform them to their linguistic roots. A widely used stemming algorithm is the well known Porter algorithm. In practice the index terms are the stems. Stemming results in common patterns (forms) for different words (having the same or similar meaning, such as singular and plural forms), and represents a special and difficult research area in Natural Language Processing and Computational Linguistics, for example.
- Compute for each document $D_i$ the number of occurrences $f_{ij}$ of each term $t_j$ in that document.
- Calculate the total $tf_j$ for each term $t_j$ as follows

$$tf_j = \sum_{i=1}^{n} f_{ij} \tag{3.11}$$

- Rank the terms in decreasing order according to $tf_j$, and exclude the very high frequency terms, over some threshold (on the grounds that they are almost always insignificant), and the very low frequency terms as well, below some threshold (on the ground that they are not much on the writer's mind). The remaining terms can be taken as document identifiers; they will be called *index terms* (or simply terms).
- Compute for the index terms $t_j$ a *significance factor* (or *weight*) $w_{ij}$ with respect to each document $D_i$. Several methods are known. One of these is, for instance, the *inverse document frequency* (*idf*) weight method, according to which the weight $w_{ij}$ is given by the following formula:

$$w_{ji} = -\log_2(df_j/n) \tag{3.12}$$

where $n$ is the total number of documents, $df_j$ is the number of documents in which term $t_j$ occurs (it can be multiplied by $f_{ij}$). Alternatively:

$$w_{ji} = f_{ij}(\log_2 n - \log_2 df_j + 1) \tag{3.13}$$

where $f_{ij}$ is the occurrence of term $t_j$ in document $D_i$.

It can be seen that $w_{ij}$ increases with $f_{ij}$, and decreases with $df_j$ (while $n$ is constant).

Another (very simple) method to compute the weights is, for example, the word frequency:

$$w_{ji} = \frac{f_{ij}}{n_i} \tag{3.14}$$

where $n_i$ denotes the number of terms in $D_i$ ('length' of $D_i$).

An object $Q_k$, called *query*, coming from a user, is also conceived as being a (much smaller) document, a vector $\mathbf{v}_k$ can be computed for it, too, in a similar way.

*Retrieval* is now defined as follows:

*A document $D_i$ is retrieved in response to a query $Q_k$, if the document and the query are "similar enough", a similarity measure $s_{ik}$ between the document (identified by $\mathbf{v}_i$) and the query (identified by $\mathbf{v}_k$) is over some threshold $K$, i.e.*

$$s_{ik} = s(\mathbf{v}_i, \mathbf{v}_k) > K \tag{3.15}$$

Typical *similarity measures* are as follows:

**a) Dot product (simple matching coefficient; inner product):**

$$s_{ik} = (\mathbf{v}_i, \mathbf{v}_k) = \sum_{j=1}^{m} w_{ji}w_{jk} \tag{3.16}$$

If $D_i$ and $Q_k$ are conceived as sets of terms, the set theoretic counterpart of the simple matching coefficient is:

$$s_{ik} = |D_i \cap Q_k|$$

**b) Cosine measure: $c_{ik}$**

$$s_{ik} = c_{ik} = \frac{(\mathbf{v}_i, \mathbf{v}_k)}{\|\mathbf{v}_i\| \cdot \|\mathbf{v}_k\|} \tag{3.17}$$

If $D_i$ and $Q_k$ are conceived as sets of terms, the set theoretic counterpart of the cosine measure is:

$$c_{ik} = \frac{|D_i \cap Q_k|}{(|D_i| \cdot |Q_k|)^{1/2}}$$

c) **Dice's coefficient:** $d_{ik}$

$$s_{ik} = d_{ik} = \frac{2 \cdot (\mathbf{v}_i, \mathbf{v}_k)}{\sum\limits_{j=1}^{m} (w_{ji} + w_{jk})} \qquad (3.18)$$

If $D_i$ and $Q_k$ are conceived as sets of terms, the set theoretic counterpart of Dice's coefficient is:

$$d_{ik} = \frac{2 \cdot |D_i \cap Q_k|}{|D_i| + |Q_k|}$$

d) **Jaccard's coefficient:** $J_{ik}$

$$s_{ik} = J_{ik} = \frac{(\mathbf{v}_i, \mathbf{v}_k)}{\sum_{j=1}^{m} \dfrac{w_{ji} + w_{jk}}{2^{w_{ji} w_{jk}}}} \qquad (3.19)$$

If $D_i$ and $Q_k$ are conceived as sets of terms, the set theoretic counterpart of Jaccard's coefficient is:

$$J_{ik} = \frac{|D_i \cap D_j|}{|D_i \cup D_j|}$$

e) **Overlap coefficient:** $O_{ik}$

$$s_{ik} = O_{ik} = \frac{(\mathbf{v}_i, \mathbf{v}_k)}{\min \left( \sum\limits_{j=1}^{m} w_{ji}, \sum\limits_{j=1}^{m} w_{jk} \right)} \qquad (3.20)$$

If $D_i$ and $Q_k$ are conceived as sets of terms the set theoretic counterpart of the overlap coefficient is:

$$O_{ik} = \frac{|D_i \cap D_j|}{\min(|D_i|, |D_j|)}$$

EXAMPLE 3.2

Let the set of documents be

$$O = \{O_1, O_2, O_3\}$$

where

$O_1$ = *Bayes' Principle: The principle that in estimating a parameter one should initially assume that each possible value has equal probability (a uniform prior distribution).*

$O_2$ = *Bayesian Decision Theory: A mathematical theory of decision making which presumes utility and probability functions, and according to which the act to be chosen is the Bayes act, the one with highest Subjective Expected Utility. If one had unlimited time and calculating power with which to make every decision, this procedure would be the best way to make any decision.*

$O_3$ = *Bayesian Epistemology: A philosophical theory which holds that the epistemic status of a proposition (i.e. how well proven or well established it is) is best measured by a probability and that the proper way to revise this probability is given by Bayesian conditionalisation or similar procedures. A Bayesian epistemologist would use probability to define, and explore the relationship between, concepts such as epistemic status, support or explanatory power.*

Let the set $T$ of terms be:

$T = \{t_1 = $ Bayes' Principle, $t_2 = $ probability, $t_3 = $ decision making, $t_4 = $ Bayesian Epistemology, $t_4 = $ Bayes$\}$

Conceiving the documents as sets of terms, they become:

$$D = \{D_1, D_2, D_3\}$$

where

$D_1$ = {Bayes' Principle, probability, Bayes}

$D_2$ = {probability, decision making, Bayes}

$D_3$ = {probability, Bayesian Epistemology}

Let the query $Q_k$ be (as a set of terms):

$Q_k$ = {probability, decision making}

Notice that, in the *VIR*, the query is not a Boolean expression anymore (as in *BIR*). Using, for example, Dice's coefficient (set theoretical form) we obtain the following:

$d_{1k}$ = 2/5, $d_{2k}$ = 4/5, $d_{3k}$ = 1/2

Using the threshold K = 0.7, for example, doucument $D_2$ is retrieved.

See Appendix 1 for another, more complex, example along with complete procedures (in MathCAD 8 Plus Professional).

As we have seen, both the query and the document are represented as *vectors* of real numbers. A similarity measure is meant to express a likeness between a query and a document.

It can be easily seen that the similarity measure typically has the following three basic properties:

– It is usually normalised, it takes on values between 0 and 1. (We shall call this property *normalization*.)
– Its value does not depend on the order in which the query and the document are considered, they are interchangeable in formulae. (We shall call this property *symmetry* or *commutativity*).
– It is maximal, equal to 1, when the query and the document are identical (exception: dot product; but notice that all the others are different normalised forms of it). (We shall call this property *reflexivity*.)

Note.
All but 3.16 are normalised and reflexive. However, 3.16 itself can be made normalised and reflexive (see Chapter 4 for more on this).

Those documents are said to be retrieved in response to a query for which the similarity measure exceeds some *threshold*.

Thus, in general, the vector *IR* can be mathematically formalised as follows:

Let $D$ be a set of elements called *documents*. A mapping

$$\sigma: D \times D \rightarrow [0, 1] \tag{3.21}$$

is called a *similarity* if the following three properties a) through c) hold:

- $0 \leq \sigma(a, b) \leq 1, \forall a, b \in D$, normalisation;
- $\sigma(a, b) = \sigma(b, a), \forall a, b \in D$, symmetry or commutativity;
- $a = b \Rightarrow \sigma(a, b) = 1, a, b \in D$, reflexivity.

Let $q \in D$ be a *query*, and $\tau \in \mathbf{R}$ be a real *threshold* value. The set $\mathfrak{R}(q)$ of *retrieved* documents in response to query $q$ is defined as follows:

$$\mathfrak{R}(q) = \{d \in D | \sigma(d, q) > \tau, \tau \in \mathbf{R}\} \tag{3.22}$$

Depending on weight values, there are two basic variants of *VIR*, namely:

*Binary VIR (BVIR)*: $w_{ij} \in \{0, 1\}$, where

- $w_{ij} = 0$ means that term $t_j$ does not occur in document $D_i$ (it is absent, or, equivalently, it does not describe/pertain to),
- $w_{ij} = 1$ means that term $t_j$ occurs in document $D_i$ (it is present, or, equivalently, it describes/pertains to).

*NonBinary VIR (NBVIR)*: $w_{ij} \in [0, 1]$.

Obviouly, the *BVIR* is a special case of *NBVIR*. See Appendix 1 for examples for both, along with complete procedures (in MathCAD 8 Plus Professional).

## 1.2.1    Bibliographical Remarks

The *vector IR (VIR)* took shape as a result of the work of, e.g., Luhn [1959], Salton [1965, 1968, 1971], Salton and McGill [1983], van Rijsbergen [1977, 1979], Forsyth [1986].

See, for example, Egghe, Rousseau and Ronald [1997], Kang and Choi [1997] for more on specific similarity functions and further detailed and specific properties.

Salton and McGill [1983] state that the process of answering a query is an interactive and iterative process, in which the retrieved documents are likely to be relevant to the user's information need (expressed by the query). Nevertheless, he admits that whether the retrieved documents are really relevant or not, is ultimately up to the user himself.

The main practical problems to be solved are: Which are the two thresholds values for excluding high and low frequency terms? Which formula to use for computing the weights $w_{ji}$? How to choose the similarity measure and the associated threshold values?

## 1.3     Probabilistic Model

The *Probabilistic* model of *IR* (*PIR*) is a classical model of *IR*. Retrieval is based on whether a probability of relevance (relative to a query) of a document is higher than that of a non–relevance (and exceeds a threshold value).

This model can be formulated in words as follows.

- Given a *query*, *documents*, and a *cut off* value.
- The *probabilities* that a document is *relevant* and *irrelevant* to the query are calculated.
- An optimal way of retrieval is as follows:

  - The documents with probabilities of relevance *at least* that of irrelevance are *ranked* in decreasing order of their relevance.
  - Those documents are retrieved whose probabilities of relevance in the ranked list *exceed* the cut off value.

A mathematical formalisation of the above is as follows. Let

- $D$ be a set of objects called *documents*,
- $Q \in D$ a *query*,
- $\alpha \in \mathbf{R}$ a *cut off value*, and
- $P(R|(Q, d))$ and $P(I|(Q, d))$ the conditional probability that object $d$ is *relevant* ($R$) and *irrelevant* ($I$), respectively, to query $Q$.

The *retrieved* documents in response to query $Q$ belong to the set $\Re(Q)$ defined as follows:

$$\Re(Q) = \{d | P(R|(Q, d)) \geq P(I|(Q, d)), P(R|(Q, d)) > \alpha, \alpha \in \mathbf{R}\} \qquad (3.23)$$

Usually the retrieved documents are first ranked and this ordered list is cut afterwards. From a mathematical point of view, the requirement of ranking the documents is contained within the set $\mathcal{R}(Q)$ (the concept of a set allows for this though the elements of a set need not be ordered).

The inequality

$$P(R|(Q, d)) \geq P(I|(Q, d)) \tag{3.24}$$

is called *Bayes' Decision Rule*.

A basic model is as follows. Given a set $D$ of elements called *documents*:

$$D = \{D_1, ..., D_i, ..., D_n\} \tag{3.25}$$

a set $T$ of elements called *index terms*:

$$T = \{t_1, ..., t_k, ..., t_m\} \tag{3.26}$$

and a set $F_i$ of non-negative integers:

$$F_i = \{f_{i1}, ..., f_{ik}, ..., f_{im}\} \tag{3.27}$$

where $f_{ik}$ represents the number of *occurrences* of term $t_k$ in document $D_i$. A *weights vector* $\mathbf{w}_i$:

$$\mathbf{w}_i = (w_{i1}, ..., w_{ik}, ..., w_{im}) \tag{3.28}$$

where $w_{ik}$ is the *weight* (significance) of term $t_k$ in the $i$th document $D_i$, is defined, for example as follows:

$$w_{ik} = \log \frac{P(f_{ik}|R)}{P(f_{ik}|I)} \tag{3.29}$$

where $P(f_{ik}|R)$ and $P(f_{ik}|I)$ denote the probability that a relevant or irrelevant (to a query), respectively, document has $f_{ik}$ occurrences of term $t_k$.

Let $h$ denote a similarity function between two vectors $\mathbf{x} = (x_1, ..., x_n)$ and $\mathbf{y} = (y_1, ..., y_n)$ defined by the dot product:

$$h(\mathbf{x}, \mathbf{y}) = (\mathbf{x}, \mathbf{y}) = \sum_{i=1}^{n} x_i y_i \tag{3.30}$$

It is assumed that (*optimal retrieval hypothesis*):

*an optimal way to retrieve documents is to retrieve them in descending order of relevance, that is, for any two documents $D_i$ and $D_j$ we have :*

$$P(R|D_i) \geq P(R|D_j) \tag{3.31}$$

where $P(\ |\ )$ denotes conditional probability (i.e. that a document is relevant).

It can be demonstrated that [see, e.g., Chapter 4 or Meng and Park, 1989]:

$$P(R|D_i) \geq P(R|D_j) \Leftrightarrow h(\mathbf{w}_q, \mathbf{w}_i) \geq h(\mathbf{w}_q, \mathbf{w}_j) \tag{3.32}$$

In order to use this model in practice, the following method can be applied.

Let $q$ denote a query, and let us look at it as if it were the set $T$. Let $|T| = N$. In order for a document $D_i$ to be retrieved in response to $q$, the following condition can be used:

$$\sum_{k=1}^{N} f_{ik} \geq K \tag{3.33}$$

where $K$ is a threshold.

The retrieved documents are then presented to the user, who judges which are relevant and which are not. This is called *relevance feedback*. From the retrieved and relevant documents the following table is constructed first for each term $t_k$ in $T$:

Table 3.1

| $T_k =$ | 0 | 1 | . . . | $j$ | . . . |
|---------|-----|-----|-------|-----|-------|
|         | $b_0$ | $b_1$ | . . . | $b_j$ | . . . |

where $T_k$ is a variable associated to term $t_k$ and takes on the values 0, 1, ...,$j$, ... (which can be interpreted as numbers of occurrences), $b_j$ is the number of relevant and retrieved documents having $j$ occurrences of term $t_k$. Then $P(f_{ik}|R)$ is given by:

$$P(f_{ik}|R) = \frac{b_j}{b_0 + b_1 + \ldots} \tag{3.34}$$

for $f_{ik} = j$. Thus, the probabilities that a relevant document $D_i$ has $j$ occurrences of the $k$th term are computed. The same method is used for the irrelevant documents, too.

Calculate new weights vectors $\mathbf{w}_i$ for documents, assign weight 1 to each query term, and use the optimal retrieval hypothesis to retrieve and rank order new documents.

Note

This method gives better results if the probabilities are (re–)computed using accumulated statistics for many queries.

EXAMPLE 3.3

Let the set of documents be

$$D = \{D_1, D_2, D_3\}$$

where

$D_1$ = *Bayes' Principle: The principle that, in estimating a parameter, one should initially assume that each possible value has equal probability (a uniform prior distribution).*

$D_2$ = *Bayesian Decision Theory: A mathematical theory of decision making which presumes utility and probability functions, and according to which the act to be chosen is the Bayes act, i.e. the one with highest Subjective Expected Utility. If one had unlimited time and calculating power with which to make every decision, this procedure would be the best way to make any decision.*

$D_3$ = *Bayesian Epistemology: A philosophical theory which holds that the epistemic status of a proposition (i.e. how well proven or well established it is) is best measured by a probability and that the proper way to revise this probability is given by Bayesian conditionalisation or similar procedures. A Bayesian epistemologist would use probability to define, and explore the relationship between, concepts such as epistemic status, support or explanatory power.*

Let the query $q$ be:

$q$ = *probability*

$T = \{t_1 = \text{probability}\}$, $k = 1$. In order to retrieve an initial set of documents, $f_{i1}$, $i = 1, 2, 3$, are calculated first: $f_{11} = 1$, $f_{21} = 1$, $f_{31} = 3$. Taking $K = 1$, documents $D_1$, $D_2$ and $D_3$ are retrieved: $\Sigma_k f_{1k} = 1$, $\Sigma_k f_{2k} = 1$, $\Sigma_k f_{3k} = 3 \geq K$. In a relevance feedback, $D_3$ is judged as relevant, whereas $D_1$ and $D_2$ as irrelevant. The probabilities of relevance are: $P(f_{i1}=1|R) = 0$, $P(f_{i1}=3|R) = 1$, and those of irrelevance are: $P(f_{i1}=1|I) = 1$, $P(f_{i1}=3|I) = 0$. The weights vectors for documents are as follows: $\mathbf{w}_1 = (-\infty)$, $\mathbf{w}_2 = (-\infty)$, $\mathbf{w}_3 = (\infty)$. The query vector is $\mathbf{w}_q = (1)$. In the retrieved rankorder $D_3$ preceeds $D_2$ and $D_1$.

See Appendix 2 for a more complex example along with complete procedures (in MathCAD 8 Plus Professional).

A special case of the *PIR* is the case when the documents are represented by binary vectors indicating the presence/absence of terms. Let $Q$ be a set of elements called *queries*:

$$Q = \{q_1, ..., q_k, ..., q_p\} \tag{3.35}$$

Let us consider the mappings:

$$f_D: D \to 2^T, \; f_D(d_j) = \mathbf{x}_j = (x_{j1}, ..., x_{ji}, ..., x_{jm}) \tag{3.36}$$

and

$$f_Q: Q \to 2^T, \; f_Q(q_k) = \mathbf{y}_k = (y_{k1}, ..., y_{ki}, ..., y_{km}) \tag{3.37}$$

where $x_{ji}$ and $y_{ki}$ are equal to 0 or 1 indicating the absence or presence of index term $t_i$ in document $D_j$ and query $q_k$, respectively.

Documents with the same binary vector description, say $\mathbf{x}$, are indistinguishable (for the system). This leads to an equivalence relation over $D$ which induces partitions, denoted by $[\mathbf{x}]_D$, over $D$. The same holds for $Q$, too (with the notations: $\mathbf{y}$ for the binary vector description; $[\mathbf{y}]_Q$ for the equivalence class).

A *relevance relationship*, denoted by $R$, between a document $d_j$ and a query $q_k$ is introduced as follows:

$$R = \{(d_j, q_k)|d_j \text{ is relevant to } q_k\} \tag{3.38}$$

It is assumed that such a relation is *a priori* known for a group of queries and documents. The relationship $R$ is hence a subset of the Cartesian product $D \times Q$, which is hence divided into two disjoint subsets.

The *IR* system is unable to predict with certainty whether a document is relevant to a query based solely on the binary vector descriptions.

Nevertheless, a measure for the uncertainty of such a prediction can be introduced.

Given a pair of binary vectors **x** and **y**, a probability $P(\mathbf{x}, \mathbf{y})$ is defined as follows:

$$P(\mathbf{x}, \mathbf{y}) = \frac{|\{(\mathbf{x}, \mathbf{y})_{D \times Q}\}|}{|\{(\mathbf{x}, \mathbf{y})\}|}, \tag{3.39}$$

where $|\{(\mathbf{x}, \mathbf{y})_{D \times Q}\}|$ is the number of document–query pairs with the same binary vector description **x** for the documents, and **y** for the queries, whilst $|\{(\mathbf{x}, \mathbf{y})\}|$ is the total number of document–query pairs. A conditional probability $P(R|(\mathbf{x}, \mathbf{y}))$ is also defined, as follows:

$$P(R|(\mathbf{x}, \mathbf{y})) = \frac{|\{(\mathbf{x}, \mathbf{y})_{D \times Q}\}| \cap R}{|\{(\mathbf{x}, \mathbf{y})_{D \times Q}\}|} \tag{3.40}$$

A document described by **x** is judged to be relevant to a query described by **y** if:

$$P(R|(\mathbf{x}, \mathbf{y})) > P(I|(\mathbf{x}, \mathbf{y})) \tag{3.41}$$

where $I$ means non–relevant and (according to Bayes' Formula):

$$P(R|(\mathbf{x}, \mathbf{y})) = \frac{P((\mathbf{x}, \mathbf{y})|R)P(R)}{P((\mathbf{x}, \mathbf{y}))} \tag{3.42}$$

$$P(I|(\mathbf{x}, \mathbf{y})) = \frac{P((\mathbf{x}, \mathbf{y})|I)P(I)}{P((\mathbf{x}, \mathbf{y}))} \tag{3.43}$$

where $P(R)$ and $P(I)$ are *a priori* probabilities of relevance and nonrelevance of a document, with the following independence conditions:

$$P((\mathbf{x}, \mathbf{y})|R) = P((x_1, y_1)|R) \cdot \ldots \cdot P(x_n, y_n|R) \tag{3.44}$$

and

$$P((\mathbf{x}, \mathbf{y})|I) = P((x_1, y_1)|I) \cdot \ldots \cdot P(x_n, y_n|I) \tag{3.45}$$

Specifically:

$$P((\mathbf{x}, \mathbf{y})|Z) = \prod_{i=1}^{n} p_{i0}^{(1-x_i)(1-y_i)} \, p_{i1}^{(1-x_i)y_i} \, p_{i2}^{x_i(1-y_i)} \, p_{i3}^{x_i y_i} \tag{3.46}$$

where

$$p_{ik} = \frac{n_{ik}}{\sum_{j=0}^{3} n_{ij}} \quad \text{for } Z = R, \quad k = 0, 1, 2, 3 \tag{3.47}$$

$$p_{ik} = \frac{m_{ik}}{\sum_{j=0}^{3} m_{ij}} \quad \text{for } Z = I, \quad k = 0, 1, 2, 3 \tag{3.48}$$

where $n_{ik}$ and $m_{ik}$ are the cooccurrence frequencies of index term $t_i$ in the document–query pair represented by $(\mathbf{x}, \mathbf{y})$ with respect to $R$ and $I$, respectively.

The retrieval condition (Bayes' Decision Rule) reduces to:

$$P(R)P((\mathbf{x}, \mathbf{y})|R) > P(I)P((\mathbf{x}, \mathbf{y})|I) \tag{3.49}$$

From this, the following function $g$, called a *discriminant function*, can be contructed:

$$g(\mathbf{x}, \mathbf{y}) = \log\,(P((\mathbf{x}, \mathbf{y})|R) \,/\, P((\mathbf{x}, \mathbf{y})|I)) + \log\,(P(R) \,/\, P(I)) \tag{3.50}$$

Thus, the retrieval condition becomes:

$$P(R)P((\mathbf{x}, \mathbf{y})|R) > P(I)P((\mathbf{x}, \mathbf{y})|I) \Leftrightarrow g(\mathbf{x}, \mathbf{y}) > 0 \tag{3.51}$$

Using (3.46), the discriminant function becomes a quadratic function on the components of $\mathbf{x}$ and $\mathbf{y}$:

$$g(\mathbf{x}, \mathbf{y}) = \sum_{i=0}^{n} (a_i x_i + b_i y_i + c_i x_i y_i) + C \tag{3.52}$$

where

$$a_i = \log \frac{p_{i2}\, q_{i0}}{p_{i0}\, q_{i2}}$$

$$b_i = \log \frac{p_{i1}\, q_{i0}}{p_{i0}\, q_{i1}}$$

$$c_i = \log \frac{p_{i0}\, p_{i3}\, q_{i1}\, q_{i2}}{p_{i1}\, p_{i2}\, q_{i0}\, q_{i3}}$$

$$C = \sum_{i=0}^{n} \log \frac{p_{i0}}{q_{i0}} + \log \frac{P(R)}{P(I)}$$

where $q_{ik}$, $k = 0, 1, 2, 3$, are equal to $p_{ik}$ for $Z = I$. The parameters $p_{ik}$ are estimated using an initially retrieved set of documents. However, historically accumulated data improve the estimations.

### 1.3.1    Bibliographical Remarks

The *PIR* model took shape as a result of the work of Maron and Kuhns [1960], Robertson and Spark–Jones [1976], Cooper and Maron [1978], van Rijsbergen [1977, 1979], Robertson, Maron and Cooper [1982], Yu, Meng and Park [1989], Wong and Yao [1990].

Different mathematical methods of computing the probabilities $P(R|(Q, d))$ and $P(I|(Q, d))$, as well as properties and applications, are investigated and applied by, e.g., Yu, Meng and Park [1989], Callan, Croft and Harding [1992], Wong, Butz and Xiang [1995], Croft, Harding and Weir [1997], Huang and Robertson [1997], Fuhr [1992], Wong and Yao [1993], van Rijsbergen [1992].

Although $P(R|(Q,d))$ and $P(I|(Q, d))$ are called the probability of relevance and irrelevance, respectively, of document $d$ to query $Q$, and their estimation is based on Bayes' Formula, they are, actually, probabilities *associated* with $d$ when it is *judged to be relevant* and *irrelevant*, respectively, to $Q$, rather than 'true' conditional probabilities in Kolmogoroff's sense [Kolmogoroff, 1950].

It is emphasised that the difficult practical problems are: i) the value of the threshold $K$ for the initially retrieved documents; ii) the number of relevant documents retrieved by a query is usually too small for the probability to be estimated accurately; to remedy this situation, statistics accumulation for numerous queries is required.

## 2.    NONCLASSICAL MODELS OF INFORMATION RETRIEVAL

The nonclassical models of *IR* appeared as a reaction to the classical models in order to enhance retrieval performance. These differ from the classical ones in that retrieval is based on principles that are different from those in the classical models, i.e., others than similarity, probability, Boolean operators.

There are several categories of nonclassical *IR* models, as follows:

– Information Logic *IR*;
– Situation Theory *IR*;
– Interaction *IR*.

The Information Logic *IR* (*ILIR*) is based on a special logical technique, called *logical imaging*, and conceives retrieval as a logical inference process from documents to query, with a degree of uncertainty of such an inference.

The Situation Theory *IR* (*STIR*) proposes an Information Calculus for *IR* based on Situation Theory. Retrieval is viewed as a flow of information from documents towards a query.

The Interaction *IR* ($I^2R$) is based on the concept of interaction as understood in the Copenhagen Interpretation of Quantum Mechanics. In this model, the query effectively interacts with the interconnected documents first, and retrieval is conceived as a result of this interaction.

## 2.1    Information Logic Model

The *Information Logic* model of *IR* (*ILIR*) represents a nonclassical approach to *IR* foundation. Retrieval is conceived as an (uncertain) inference from a document to query.

## 2.1.1    Basic Concepts

### 2.1.1.1    Document

The document, denoted by $d$, is thought of as a chunk of coherent text, e.g., a finite set $S$ of sentences. These sentences can constitute an abstract, an article, a book, etc., or they may be 'about it' rather than 'in it', i.e., they represent the description of a domain (of an article, for instance, rather than the article itself).

### 2.1.1.2    Query

A user is supposed to express his/her information need by formulating a query, denoted by $q$, which is also a piece of text, but usually without the complexity of a document.

### 2.1.1.3    Retrieval

Using the query $q$, a document (or documents) $d$ is retrieved somehow. This 'somehow' may be thought of as a sort of inference process from $S$ to $q$, denoted as follows:

$$S \to q \tag{3.53}$$

This is not the usual material implication which is true in all cases except when the premise is true and the conclusion is false. In deciding whether to retrieve a document $d$, the inference $S \to q$ should be evaluated somehow. Therefore, a measure of uncertainty of this inference is considered, and denoted by $P(S \to q)$. Further, the following principle is formulated and used:

**Van Rijsbergen's Principle**

*Given any two sentences $x$ and $y$. A measure of the uncertainty of $y \to$ $x$ relative to a given data set is determined by the minimal extent to which one has to add information to the data set in order to establish the truth of $y \to x$.*

With this principle, the notion of *minimal extra information* is introduced, which together with $y$ yields to $x$ (via the use of Modus Ponens).

## 2.1.2    Information Retrieval

The information logic model of *IR* is described using a special technique from logic called *logical imaging*.

Let $d$ be a document, $s$ a sentence, $q$ a query, and $d_s$ the document consisting of $d$ plus the minimal extra information for $s$ to be true in $d_s$. The inference $s \to q$ is supposed to be true relative to $d$ if and only if $q$ is true relative to $d_s$.

Documents are assigned probabilities so that they sum to unity over documents:

$$\sum_d P(d) = 1 \tag{3.54}$$

Let us introduce the following function:

$$d(s) = 1, \text{ if } s \text{ is true at } d \tag{3.55}$$

$$d(s) = 0, \text{ if } s \text{ is false at } d \tag{3.56}$$

Then:

$$d(s \to q) = d_s(q) \tag{3.57}$$

Let us define the probability $P$ of a sentence $s$ over a set of some documents $d$ as follows:

$$P(s) = \sum_d P(d)d(s) \tag{3.58}$$

Let us define another probability, $P'$, which 'shifts' the probability from the not–$y$ documents to the $y$ documents, i.e. to $d$ together with the extra minimal information, as follows (we transfer the probability of $d$ to its $d_s$, the most similar document to $d$ where $s$ is true):

$$P'(d') = \sum_d P(d)u(d') \tag{3.59}$$

where

$$u = 1, \text{ if } d' = d_s \tag{3.60}$$

$$u = 0, \text{ if } d' \neq d_s \tag{3.61}$$

Then one can write (the 'index' of the equal sign is the number of the formula according to which the equality can be written):

$$P'(q) =_{3.58} \Sigma_{d'} P(d')d'q)$$

$$=_{3.59} \Sigma_{d'} (d'(q) \Sigma_d P(d)u)$$

$$= \Sigma_d (\Sigma_{d'} d'(q)(P(d)u))$$

$$= \Sigma_d (P(d) \Sigma_{d'} d'(q)u) \tag{3.62}$$

As we are interested in a case when $q$ becomes true relative to $d$, we set $d' = d_s$, and:

$$\Sigma_d P(d)d_s(q) = \Sigma_d P(d)d(s \rightarrow q) = P(s \rightarrow q) \tag{3.63}$$

The relation

$$P'(q) = P(s \rightarrow q) \tag{3.64}$$

suggests that — relative to a data set — answering a query $q$ means a process of inference from $s$ to $q$ with a measure of uncertainty of this inference. In other words, a query should be derived, inferred from documents somehow.

Note
Formally, a simpler derivation of the result that a measure of the inference $s \rightarrow q$ is equal to a measure of $q$, is as follows. Taking s := $s \rightarrow q$ in 3.58, and using 3.57 we have: $P(s \rightarrow q) = \Sigma_d P(d)d(s \rightarrow q) = \Sigma_d P(d)d_s(q)$. Based on 3.59, this can be re-written as $P'(d_s)$ which re-writes as some $P''(q)$, because $d_s$ may be considered as depending on $q$ ($q$ is true at $d_s$).

EXAMPLE 3.4
1) Consider the following two documents (from ADI test collection):

$D_1$ = *An English–like system of Mathematical Logic is a formally defined set of sentences whose vocabulary and grammar resemble English, with an algorithm which translates any sentence of the set into a notation for*

*Mathematical Logic. The use of this system for content retrieval is described.*

$D_2$ = *A new centralised information retrieval system for the petroleum industry, including a computer search system, is proposed.*

and the following query:

$Q$ = *The use of abstract mathematics in information retrieval.*

Taking document $D_2$, there is nothing that would indicate the use of any abstract mathematics. Hence, in principle, $D_2$ should be expanded with a potentially infinite quantity of extra information in order for it to lead to $Q$, even though the 'computer search system' can be viewed as a practical equivalent of 'information retrieval'. Taking document $D_1$, the use of a 'formal definition' based on 'Mathematical Logic' pertains to the meaning of the notion of 'abstract mathematics', and this technique is applied to 'content retrieval' which is a form of 'information retrieval'. Hence, for document $D_1$, the amount of extra information to be added, in order to to establish the truth of the query, is much less than that for $D_2$. Hence, $D_1$ is preferred to $D_2$, and thus retrieved.

2) Another — very special example — is represented by an axioms–based formal system. In this case, many queries (beware: Gödel's Incompleteness Theorem) can be formally deduced from the axioms of the system. A measure of the uncertainty of such a deduction is proportional to the 'amount' of extra information, e.g., number of intermediate theorems, lemmas, etc., which should be considered to make our query true (or false). This 'amount' of exta–information is, at the same time, a measure of how 'far' the query is from the axioms. Thus theorem proving, for example, is a particular case of Information Logic *IR*.

## 2.2　　Situation Theory Model

The model of *IR* based on Situation Theory [Devlin, 1991] is a nonclassical model of *IR* (*STIR*), and constitutes a new theoretical framework. It uses van Rijsbergen's Principle, and conceives retrieval as a flow of information from documents to query.

### 2.2.1 Basic Concepts

The concept of *information* is taken as a basic concept. *Cognitive agents* (e.g., human beings, measuring apparatus) are able to perceive information. A *situation* means that different agents can pick up different information from the same source depending on the *constraints* that the agents are aware of.

For example, consider the following situation. If a student (cognitive agent) is aware of the relation (constraint) between the area $A$ of a circle and its radius $r$, $A = \pi r^2$, and is given the area of a circle, then he/she should be able to calculate the radius (information picked up) of that circle.

#### 2.2.1.1 Infon

The basic concept to describe situations and model information flow is that of an *infon* which is defined as follows:

An *infon*, denoted by $\iota$, is a structure $\iota = \langle\langle R, a_1, ..., a_n; i \rangle\rangle$, where $R$ is an $n$–ary ($n$–place) relation (between $a_1, ..., a_n$) and $i$ indicates that the relation $R$ holds ($i = 1$, the infon carries *positive information*) or does not hold ($i = 0$, the infon carries *negative information*). $i$ is called the *polarity* of the infon.

For example, the information in the sentence "Mary plays the piano." is expressed by the following infon $\iota = \langle\langle plays, Mary, piano; 1 \rangle\rangle$.

Thus information is not represented by a truth value anymore, but by an infon.

#### 2.2.1.2 Support

Whether an infon is carrying positive or negative information depends on the situation that "makes it true" — this is modelled by the notion of a *support*:

The *support* relation between a situation $s$ and an infon $\iota$ is denoted by $s \models \iota$ and means that situation $s$ makes infon $\iota$ true.

For example, the infon $\iota = \langle\langle plays, Mary, piano; 1 \rangle\rangle$ is made true by a situation $s_1 = $ "I see Mary playing the piano." or $s_2 = $ "I can here somebody playing the piano, and I know this can only be Mary since she is the only person who can play a piano.".

### 2.2.1.3    Type

By parameterising situations $s_j$ and elements $a_i$, it is possible to express a common information. This is captured in the notion of a type:

A common information expressed in different infons is called a *type*.

For example, $\varphi = [\dot{c} \models \langle\langle plays, Mary, \dot{g}; 1\rangle\rangle]$ is a type expressing the common information "Mary plays". $\dot{g}$ is a parameter meaning an instrument, and $\dot{c}$ is another parameter denoting a situation.

Instantiating the parameters is called *anchoring*. Any situation supporting an anchored infon is said to be of the respective type $\varphi$; this is denoted by $s \models \varphi$.

A type can be *compound* if it consists of several infons.

### 2.2.1.4    Constraint

The flow of information is modelled by the concept of a constraint.

A *constraint*, denoted by $\rightarrow$, is a link (an informational connection or relationship) between two types $\varphi_1$ and $\varphi_2$: $\varphi_1 \rightarrow \varphi_2$. The constraint means that a situation $s_1$ of type $\varphi_1$ carries the information that a situation $s_2$ is of type $\varphi_2$.

Formally:

$$(\varphi_1 \rightarrow \varphi_2) \Leftrightarrow ((s_1 \models \varphi_1) \Rightarrow (s_2 \models \varphi_2)) \tag{3.65}$$

For example, the types $\varphi_1 = [\dot{c} \mid \dot{c} \models \langle\langle plays, Mary, \dot{g}; 1\rangle\rangle]$ and $\varphi_2 = [\dot{c} \mid \dot{c} \models \langle\langle here, music, \dot{g}; 1\rangle\rangle]$ are linked by the fact that if Mary plays an instrument then a person, listening to her playing, can hear the music.

### 2.2.1.5    Channel

While a constraint is a relationship between types, a channel is a link between situations [Barwise, 1993].

A *channel*, denoted by $c$, is a link between two situations $s_1$ and $s_2$ denoted by $s_1 \mapsto_c s_2$, and means that $s_1$ contains information about $s_2$. $s_1$ is called the *signal situation* and $s_2$ the *target situation*.

## 2.2.2  Information Retrieval

The information contained in an object document can be thought of as a situation $d$. The information need expressed by an object query can be modelled by a type $\varphi$.

If

$$d \models \varphi \tag{3.66}$$

then the document $d$ is said to be relevant to query $\varphi$.

However, if the document does not support the query this does not necessarily mean that the document is not relevant to the query; it may be that there is not enough explicit information to show relevance. To remedy this, information flows are used:

*the relevance of a document d to a query $\varphi$ is determined by the information flow necessary to transform d into d' such that d' $\models \varphi$.*

This transformation can be conceived as a flow of information between situations, i.e., channels between documents. For example, a document $d$ can be transformed into another document $d'$ using the concepts and relationships of an appropriate thesaurus. The semantic relationships in the thesaurus are the channels: e.g., synonyms, broader than, narrower than, related.

Alternatively, a somewhat different model of *IR* based on Situation Theory [Devlin, 1991] is described in [Huibers and Bruza, 1996, van Rijsbergen and Lalmas, 1996]. This model is defined as follows. Given a set

$$D = \{d_k | \ k = 1, 2, ..., M\} \tag{3.67}$$

of *documents*. Let

$$S = \{s_k | \ k = 1, 2, ..., M\} \tag{3.68}$$

be an associated set of *situations*, where the situation

$$s_k = \{i_{kp} | \ i_{kp} = \langle\langle R_{kp}, a_{k1}, ..., a_{kj}, ..., a_{kn_p}, I_{kp}\rangle\rangle, \ j = 1, 2, ..., n_p, \ p = 1, ..., p_k\} \tag{3.69}$$

for $k = 1, 2, ..., M$, is a set of *infons*.

A counterpart of the concept of retrieval is represented by the concept of *situation aboutness*. This is denoted by the symbol $\mapsto$ and defined as follows:

$$s_k \mapsto s_i \Leftrightarrow \exists\, i_r \in s_i\, [s_k \models i_r] \tag{3.70}$$

Depending on how $\models$ is defined, particular *IR* models can be obtained. For example, a Coordination Level Matching (*CLM*) model is obtained if

$$s_k \mapsto s_i \Leftrightarrow s_k \cap s_j \neq \varnothing \tag{3.71}$$

### 2.2.3    Bibliographical Remarks

All the logic models of *IR* are based on van Rijsbergen's principle — which postulates that retrieval is an inference process from document to query with a degree of uncertainty of this inference — instead of a search process (as in the classical models). The origins of this model type can be found in [van Rijsbergen, 1986a, 1986b, 1989] which appeared from the recognition that the previous models of information retrieval (Boolean, vector and probabilistic) were not able to enhance effectiveness and this could only be expected from radically new *IR* models that could deal with meaning. Thus the Information Logic was suggested.

In [Watters, 1989] the formalism of predicate calculus is used to model the inference process, and a specific logic *IR* model is elaborated based on deductive rules (expert *IR* system). The classical Boolean and vector *IR* models are also expressed within this framework.

Meanings of conditionals in the probabilistic *IR* model are analysed in [van Rijsbergen, 1992] and replaced with the more general belief functions (Dempster—Schafer theory of evidence) thus offering an epistemic view of the concept of probability in *IR*. Also, the interpretation of the meaning of conditionals is shifted towards logical inference.

A semantic interpretation of the similarity measure used in the vector *IR* model is suggested in [da Silva and Milidiu, 1993]. Belief functions (from Schafer's theory of evidence) are used to represent documents: $Bel_d(t)$ represents a degree of belief in term $t$ as being a semantic representative of document $d$ (similarly for queries). Retrieval is viewed as an agreement between document and query: the probability that they are not independent.

The logical technique, called imaging, suggested for information logic is refined in [Crestani and van Rijsbergen, 1995a, 1996].

New theories of information and information flow, Situation Theory [Devlin, 1991] and Channel Theory [Barwise, 1993], are used to model retrieval in [van Rijsbergen and Lalmas, 1996; Huibers and Bruza, 1996]:

documents are situations, queries are types and retreival is a flow of information from documents to queries.

Van Rijsbergen's principle was extended and adopted in e.g., [Nie, 1989, 1992; Nie and Chiaramella, 1990] by considering an inference from the query to document, too (not just from document to query, as was initially proposed). Thus, a measure to retrieve a document depends on both an inference from document to query and from query to document. Both inferences are based on a knowledge base, and bear a degree of uncertainty.

The extended principle of minimal extra information was adopted in [Fuhr and Roelleke 1998] for hypermedia retrieval using database attribute values. Their *IR* system, HySpirit, is a practical attempt at implementing the logic *IR* model. The degree of the uncertain inference is calculated using conditionals as follows: $P(d \rightarrow q) = \Sigma_t P(q|t)P(d|t)(P(t)/P(d))$.

## 2.3 Interaction Model

In the classical philosophy of *IR* there are the following entities [see, e.g., Meadow, 1988]: document, query, relevance. The documents form an isolated and static 'world of documents' in which they are either not consciously and explictly interconnected in a flexible way, or, if they are, the interconnections are not used. The query is not a part of this 'world of documents' nor has it any influence on its structure (if any). Retrieval means a search action in the 'world of documents' initiated by the query without any effect upon the 'world of documents'. Thus, a parallel can be drawn between the classical philosophy of *IR* and the measurement process in classical physics. In the latter, a process of *measurement* is performed using a *measuring apparatus* on an *observable* (parameter of a ) system, and the measuring apparatus does not interact with the observed system at all or, equivalently, this interaction can always be reduced to zero. The corresponding elements of such a parallel are as follows (Figure 3.1):

|  |  |
|---|---|
| *query* | *measuring apparatus* |
| *documents* | *observed system* |
| *retrieval* | *measurement* |

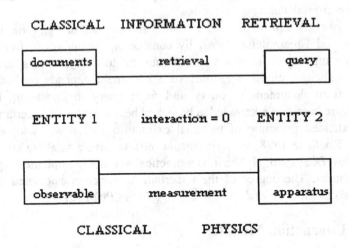

*Figure 3.1.* Parallel between classical IR and measurement in classical physics

This parallel brings attention to the other view of measurement in physics which is different from the classical view in that the interaction between apparatus and observable cannot be reduced to zero and the measured value is a result of (or is not independant) of interaction. This view is taken in Quantum Mechanics and known as the Copenhagen Interpretation of measurement in physics [von Neumann, 1927; Bohr, 1928, 1958; Heisenberg, 1971; Roland, 1994]. Thus, the following question arose: Would it be possible to make another parallel, too, namely an *IR* model similar to the Copenhagen Interpretation? (Figure 3.2)

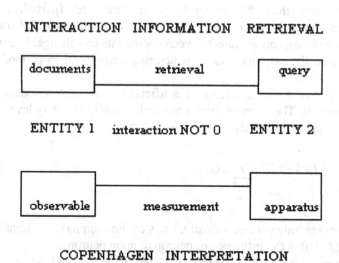

Figure 3.2. Copenhagen Interpretation and Interaction–based IR

### 2.3.1 Qualitative Model

The following question arises: Is it possible to conceive an *IR* model based on the Copenhagen Interpretation? Obviously in such a paradigm for *IR*, which may be called *I*nteraction *IR* (or briefly *I*nteraction $IR = I^2R$), the concept of an interaction is the keypoint:

– The query should interact with documents,
– Retrieval should be a result of this interaction.

1. One way in which to achieve interaction is to conceive the 'world of documents' as documents (or objects, in general) no longer being isolated, instead as being effectively and flexibly interconnected with each other, and the query, before being answered, becoming a member of this 'world of documents' (partially) re–structuring it (interaction).

For a mathematical modelling the fundamental state equation from *ANN* proves useful. Thus the 'world of documents' is conceived as being a neural network, with each document being an artificial neuron which can exhibit different levels of activity. The query is also viewed as being an artificial neuron which is integrated into the network (as if it were just another document), and thus the network is re–structured (partially): new connections appear between query and the rest of documents, and some of the already existing connections between documents can change. Integrating the query into the network and re–structuring correspond to or model the concept of interaction.

Let $o_i$, $i \in N = \{1, ..., n\}$, be an artificial neuron corresponding to an object document. The neurons form a network. Then the activity level $z_i(t)$ is given by the general network equation as follows:

$$\frac{dz_i(t)}{dt} = I_i(t) - z_i(t) + \sum_{\substack{r \in R \\ r \neq i}} f_r(z_r(t)) \tag{3.72}$$

where $t$ denotes time (or equivalently, steps of functioning), $I_i$ external input to $o_i$, and $f_r(z_r(t))$ is the influence of neuron $o_r$ upon neuron $o_i$.

The query $q$ becomes a member of the network, and let $o_q$ be the corresponding neuron. Obviously, on the one hand, some neurons $o_j$ will have an influence on $o_q$, i.e., one can write:

$$\frac{dz_q(t)}{dt} = I_q(t) - z_q(t) + \sum_{\substack{j \in Q \\ j \neq q}} f_j(z_j(t)) \tag{3.73}$$

and, on the other hand, some neurons $o_k$ will be influenced by query $o_q$, this is reflected in:

$$\frac{dz_k(t)}{dt} = I_k(t) - z_k(t) + \sum_{\substack{s \in S \\ s \neq k}} f_s(z_s(t)) \tag{3.74}$$

where $\exists\, s \in S$ such that $s = q$. (Figure 3.3)

Interaction is reflected in the summed terms of 3.73 and 3.74:

$$\sum_{\substack{j \in Q \\ j \neq q}} f_j(z_j(t)) = -I_q(t) + z_q(t) + \frac{dz_q(t)}{dt} \tag{3.75}$$

$$\sum_{\substack{s \in S \\ s \neq k}} f_s(z_s(t)) = -I_k(t) + z_k(t) + \frac{dz_k(t)}{dt} \tag{3.76}$$

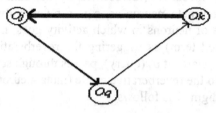

*connection between Ok and Oj before interaction*

*interaction between query Oq and other objects:*
*new connections are born and old ones can change*

*Figure 3.3* Interaction between query and documents in Interaction IR

and a measure of it is obtained by adding them up:

$$\sum_{\substack{j \in Q \\ j \neq q}} f_j(z_j(t)) + \sum_{\substack{s \in S \\ s \neq k}} f_s(z_s(t))$$

$$= -I_q(t) + z_q(t) + \frac{dz_q(t)}{dt} - I_k(t) + z_k(t) + \frac{dz_k(t)}{dt} \qquad (3.77)$$

Making the change of variables:

$$u(t) = z_k(t) + z_q(t), \quad I(t) = I_q(t) + I_k(t) \qquad (3.78)$$

and after appropriately renumbering $f_s$ and $f_j$ one gets the equation:

$$\frac{du(t)}{dt} = I(t) - u(t) + \sum_{i \in S \cup Q} f_i(z_i(t)) \qquad (3.79)$$

Thus equation (3.79) is an expression of interaction. Formally it is of the same type as the fundamental state equation.

2. Since retrieval should be a result of (or should be determined by) interaction, one possibility of conceiving retrieval is given by the following interpretation of the quatities involved in equation (3.79).

The quantities $u = z_k + z_q$ and $I = I_q + I_k$ may be viewed as being the activity level of and input to a meta–neuron whose activity level is influenced by activity levels $z_i$ (represented by the summed term). This suggests conceiving the meta–neuron as, e.g., a reverberative circle, i.e., a self–stimulating circle of neurons in which activity goes in a circle, whilst the influences (summed term) as triggering the reverberative circle, i.e. a spreading of activation starts at $o_q$ (query), passes through several neurons $o_i$ and finally gives rise to the reverberative circle (meta–neuron). (Figure 3.4)

Thus, the $I^2R$ paradigm is as follows:

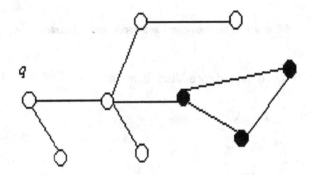

reverberative circle

*Figure 3.4*. Retrieval in Interaction IR as a result of interaction between query and interconnected documents: reverberative circles (short–term memories) triggered by a spreading of activation started at query

Given an interconnected network of object documents and a query. Retrieval is a result of an interaction between the query and documents, i.e.:

- Firstly, the query is incorporated into (becomes a member of) this network as if it were just another document. This causes partial restructuring of the network.
- Secondly, a spreading of activation starts at neuron query and spreads over from neuron to neuron until a reverberative circle (or more circles) is (are) triggered.
- Thirdly, those object documents are said to be retrieved in response to query whose corresponding neurons are members of reverberative circles.

### 2.3.2 Quantitative Model

In order to obtain an implementable model of the $I^2R$ paradigm, the following problems should be solved:

- How to interconnect the objects?
- How is the spreading of activation performed?

As we have seen, the activity level $z_i$ is given by:

$$\frac{dz_i(t)}{dt} = I_i(t) - z_i(t) + \sum_j f_j(z_j(t)) \tag{3.80}$$

where $z_i$ denotes activity level, $I_i(t)$ denotes the external input, $t$ denotes time (or steps) , $f_j(z_j(t))$ denotes the influence of $o_j$'s activity level on $z_i$. Because there is no external input we take $I_i(t) = 0$. The term $f_j(z_j(t))$ can be taken as a weight $T_j$ of the connection between $o_i$ and $o_j$ and represents the influence previously mentioned. Thus the network equation becomes:

$$\frac{dz_i(t)}{dt} = -z_i(t) + \sum_j T_j \tag{3.81}$$

The weights $T_j$ remain constant during the spreading of activation. An activation is started and spread from $o_q$ at time $t_0$ by clamping its state to 1, $z_q(t_0) = 1$, and $z_i(t_0) = 0$ $\forall i \neq q$. At time (or equivalently step) $t_1 > t_0$, $z_i$ is given by the solution of the following Cauchy–problem:

$$\frac{dz_i(t)}{dt} = -z_i(t) + \sum_k T_k , \quad z_i(t_0) = 0 \tag{3.82}$$

Denoting the sum by $s_i$, the solution is

$$z_i(t) = s_i\, e^{-t}\, (e^t - 1) \tag{3.83}$$

Thus the questions raised earlier can be answered as follows: The objects are interconnected through weighted links. When the query is interconnected all weights are re–adjusted accordingly (which is a form of learning).

In order to apply the WTA (winner takes all) strategy in practice, we can proceed as follows. Assume that object $o$ has the maximum activity $z$ of all its 'competitors' $o_i$. Then we have:

$$z(t) \geq z_i(t) \iff (s - s_i)\, e^{-t}\, (e^t - 1) \geq 0 \tag{3.84}$$

Because $e^{-t}(e^t - 1)$ is positive, it follows that $s \geq s_i$. In other words, maximum activity is equivalent to maximum weights sum

$$\sum_j T_j \tag{3.85}$$

EXAMPLE 3.5
Each document (object) $o_i$ is associated a vector $\mathbf{t}_i = (t_{ik})$, $k = 1, ..., n_i$, of identifiers. There are two pairs of links per direction. The one is the frequency of a term given a document, i.e. the ratio between the number $f_{ijp}$ of occurrences of term $t_{jp}$ in object $o_i$, and the length $n_i$ of $o_i$, i.e. total number of terms in $o_i$:

$$w_{ijp} = \frac{f_{ijp}}{n_i}, \quad p = 1, ..., n_j$$

The other is the extent to which a given term reflects the content of a document, i.e. the inverse document frequency. $f_{ikj}$ denotes the number of occurrences of term $t_{ik}$ in $o_j$, $df_{ik}$ is the number of documents in which $t_{ik}$ occurs, $w_{ikj}$ is given by the inverse document frequency formula, and thus represents the extent to which $t_{ik}$ reflects the content of $o_j$:

$$w_{ikj} = f_{ikj} \log \frac{2M}{df_{ik}}$$

(where $M = n$). The other two connections — in the opposite direction — have the same meaning as above: $w_{jik}$ corresponds to $w_{ijp}$, while $w_{jpi}$

corresponds to $w_{ikj}$. Input $s_i$ of $o_i$ is defined as the sum of the corresponding weights, as follows:

$$\sum_{p=1}^{n_j} w_{jpi} + \sum_{k=1}^{n_i} w_{jik}$$

The spreading of activation should obey a rule. A winner takes all (WTA) strategy is a good choice (Figure 3.5). Thus the activation spreads over to the most active neuron until reverberative circles are triggered when, in a practical implementation, activation spreading should be stopped or else it would circle forever.

Given two documents (objects) $o_1$ and $o_2$ as follows:

$o_1$ = *The paper is a good survey of free–text retrieval in the vector model.*
$o_2$ = *These notions are important in modelling information retrieval. They help to clearly understand the vector model of retrieval.*

The objects are now represented as tuples (collections) of terms (index terms or terms considered to reflect their content), e.g.

$o_1 = (t_{11} = $ free–text, $t_{12} = $ retrieval, $t_{13} = $ vector model),
$o_2 = (t_{21} = $ information retrieval, $t_{22} = $ vector model, $t_{23} = $ retrieval).

The associated vectors are:

$\mathbf{t}_i = (t_{i1}, \ldots, t_{ik}, \ldots, tin_i)$
$\mathbf{t}_j = (t_{j1}, \ldots, t_{jp}, \ldots, tjn_j)$

where $t_{ik}$ is the $k$th term of object $o_i$ which has $n_i$ number of terms (Figure 3.6).

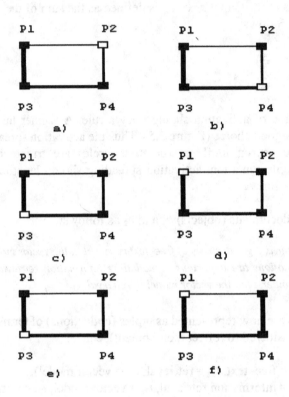

*Figure 3.5.* Interaction *IR*. Objects are interconnceted. White objects are active, grey ones are inactive. An activation is started at $P_2=Q$ (see a)) and spread through the widest connection (see b)–e)) until $P_1$ and $P_2$ activate each other in circle (see f)).

Thus in our example the connections are shown in Figure 3.7. Let us consider now a query $Q$ as follows:

$Q = $ *Usage of free text technique and user model in the vector model of etrieval.*

$Q$ is incorporated first into the interconnected objects. $Q$ becomes a member of this structure as it were just another object, say $o_3$:

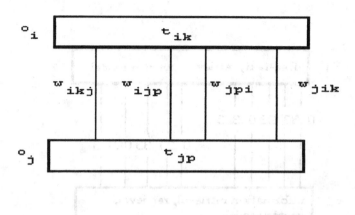

*Figure 3.6* Connections and their weights in interaction *IR* (see text).

$$Q = o_3 = (t_{31} = \text{free-text}, t_{32} = \text{user model}, t_{33} = \text{retrieval})$$

Thus, the new structure is as follows (notice that some of the old weights change; Figure 3.8). An activation is started at object $o_3$ which corresponds to query $Q$, and it operates in steps. The activation is spread towards the other two objects $o_1$ and $o_2$. Object $o_1$ is activated by the value $.33 + .33 + 0 + .47 + 0 + .3 = 1.43$, whilst object $o_2$ by $0 + .3 + 0 + 0 + 0 + .3 = .6$. Because $o_1$ $(1.43 > 0.6)$ gets the highest activity, $o_1$ is selected in the next step and becomes active whilst all the others, i.e., $o_3$ and $o_2$ will get passive. Activity is spread now from $o_1$ towards $o_2$ and $o_3$. $o_2$ is activated by $0 + .33 + .33 + 0 + .3 + .47 = 1.43$ whilst $o_3$ by $.33 + 0 + .33 + .47 + .3 + 0 = 1.43$. Thus two circles have been reached, and activation spreading is stopped. The retrieved object in response to query $Q$ are $o_1$ and $o_2$.

See a complete example, along with procedures, in *MathCAD*, for an $I^2R$ in Appendix 3.

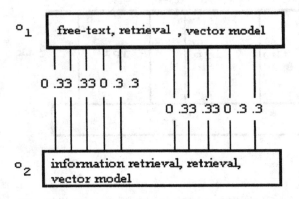

Figure 3.7. Connections (see Example in text).

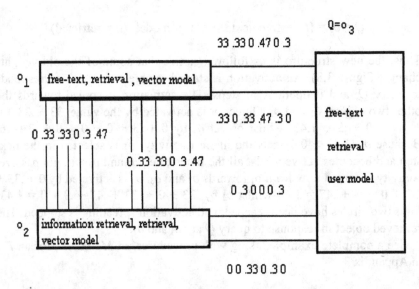

Figure 3.8. The query interacts with objects: it is interconnected with objects, and some of the old weights change. There as many links (represented by the numeric values) between $o_3$ and the other two objects as the number of weights, albeit just one line is shown.

## Notes

Obviously, there may be other choices as to activation spreading, weights calculations, and definition of interaction and retrieval. These may constitute topics for further research.

### 2.3.3 Bibliographical Remarks

The Interaction based paradigm of *IR* ($I^2R$), as well as implementations, was first suggested in [Dominich, 1992, 1993a, 1993b, 1994; van Rijsbergen, 1996] based on the concept of interaction according to the Copenhagen Interpretation in Quantum Mechanics.

The idea of flexible, multiple and mutual interconnections from $I^2R$ also appear and are investigated in [Salton, Allan and Singhal, 1996; Salton, Singhal, Mitra and Buckley, 1997]. A text document is decomposed into 'themes' or text parts which are interconnected and weighted. The retrieved items are effectively connected, and the connections change when new items are added.

The same idea of flexible interconnections is used in [Pearce and Nicholas, 1996]. There are multiple links between objects. Retrieval, however, is based on multiple look–up functions (similarities).

Another analogy between basic entities of *IR* and Quantum Mechanics [von Neumann, 1927] is suggested in [van Rijsbergen, 1996]. Documents correspond to state vectors, terms to projection operators, whilst queries to Hermitian operators. Retrieval is thought of as an interaction, applying an observable to state vectors.

The idea of flexible interconnections is suggested to be applied in multimedia *IR* in [Carrick and Watters, 1997]. Items of different nature — text, photo and sound — should be flexibly and effectively interconnected and weighted.

Liu [Liu, 1997] introduces semantic matrices to model multiple interconnections. A document is assigned a semantic matrix instead of a single vector.

Artificial Neural Networks (*ANN*) techniques are used in [Lin, 1997], namely Kohonen maps, to model a document structure which is reorganised with a query.

The idea from $I^2R$ of conceiving retrieval as a recalled memory also appears and is investigated in [Mock and Vemuri, 1997]. Retrieval is defined as recalled memory from an interrelated concept hierarchy.

The $I^2R$ model as well as implementation, together with other properties (high precision and potential use in automatising relevance feedback) are further investigated in [Dominich, 1997a, 1997b, 1998].

In [Dominich, 1999b, 2000a, 2000b], it is shown that $I^2R$ can play the role of a superstructure for the classical *IR* models in that these can be conceived as special cases of $I^2R$.

# 3.    ALTERNATIVE MODELS OF INFORMATION RETRIEVAL

The alternative models of *IR* are enhancements of the classical models, on which they rely, making use of specific techniques from different other fields.

## 3.1    Cluster Model

From a purely formal mathematical point of view the *Cluster IR* is equivalent to the *VIR* and it represents a vector space approach for reasons that will become obvious in the following. Namely, the cluster representative of the *Cluster IR* may be conceived as being a document vector in the *VIR*, or, equivalently, a document vector in the *VIR* may be viewed as a cluster representative in the cluster *IR* (representing the one–element cluster consisting of the document itself). However, because of specific (namely: clustering) techniques used to group documents, the cluster *IR* may be viewed as a — technically but not principially — different, alternative model for *IR*.

Since *IR* systems deal with large masses of data, it is useful to introduce an organisation of the stored data in order to ease access to them. One way of providing order among a collection of stored data items is to introduce a classification (or hierarchy) among them: to establish groups of related or similar items (based on different criteria). This classification is usually referred to as *clustering*.

Let

$$O = \{o_1, ..., o_n\} \tag{3.86}$$

be a set of elements called *objects*. The problem is to cluster *O*'s objects into classes. Different methods are known to cluster the set *O*. A basic method [see Salton and McGill, 1983], based on a (symmetric) *similarity matrix* and on a *threshold*, will be described in the following. Let

$$W = (w_{ij})_{n,n} \tag{3.87}$$

be a *similarity matrix*, in which $w_{ij}$ denotes a similarity measure between $o_i$ and $o_j$. Let *T* be a *threshold* value, e.g., one of the $w_{ij}$. Any pair of objects $o_i$ and $o_j$ for which $(i \neq j)$

$$w_{ij} \geq T \tag{3.88}$$

is grouped to form a cluster. The rest of the objects represent a cluster each. So, after possibly re–numbering the objects, one can write:

$$K = \{k_1, ..., k_p, ..., k_m\} \tag{3.89}$$

where

$$k_1 = \{(o_i, o_j)|\ w_{ij} \geq T\} \tag{3.90}$$

$$k_p = \{o_p|\ p \neq i, j\} \tag{3.91}$$

When the set of objects is the set of documents, then document clusters are established so that, when retrieving a document, every other document in that cluster can be retrieved and provided in response to a query. Let

$$D = \{D_1, ..., D_i, ..., D_n\} \tag{3.92}$$

be the set of elements called *documents*,

$$KD = \{k_1, ..., k_j, ..., k_m\} \tag{3.93}$$

be a set of *document clusters*,

$$T = \{t_1, ..., t_s, ..., t_t\} \tag{3.94}$$

be a set of elements called *terms* (e.g., index terms, thesaurus terms), and

$$KT = \{c_1, ..., c_p, ..., c_q\} \tag{3.95}$$

be a set of *term clusters*.
  The following binary matrix can be constructed:

$$G = (g_{sp})_{t,q} \tag{3.96}$$

where

$$g_{sp} = 1, \text{ if } t_s \in c_p \tag{3.97}$$

$$g_{sp} = 0, \text{ if } t_s \notin c_p \tag{3.98}$$

Documents $D_i$ can be assigned the following vectors $\mathbf{v}_i$:

$$\mathbf{v}_i = (w_{i1}, \ldots, w_{is}, \ldots, w_{it}; b_{i1}, \ldots, b_{ip}, \ldots, b_{iq}) \tag{3.99}$$

where
- $w_{is}$ represents the extent to which term $t_s$ pertains to document $D_i$ (weight),
- $b_{ip}$ is set to 1 if document $D_i$ pertains to class $c_p$ (i.e., $g_{sp} = 1$ for $w_{is} \neq 0$).

For documents $D_i$ belonging to the same class $k_j$, a so called *centroid* is computed, which is a representative vector $\mathbf{r}_j$ of that class (cluster), e.g., as follows:

$$\mathbf{r}_j = (a_{j1}, \ldots, a_{js}, \ldots, a_{jt}) \tag{3.100}$$

where

$$a_{js} = S^{-1} \sum_{D_i \in K_j} w_{is} \tag{3.101}$$

where $S$ represents the number of documents in class $k_j$, $w_{is}$ is the weight of term $t_s$ in document $D_i$ in class $k_j$. The summation is taken over all documents in the class $k_j$.

A search query $Q$ can also be represented as a vector as follows:

$$\mathbf{q} = (q_1, \ldots, q_s, \ldots, q_t; b_1, \ldots, b_p, \ldots, b_q) \tag{3.102}$$

where $q_s$ represents the extent to which term $t_s$ can be assigned to query $Q$ (weight), and $b_p$ is set to 1 if the query $Q$ pertains to class $c_p$ ($g_{sp} = 1$ for $q_s \neq 0$).

In the retrieval process, the query is 'compared' with the clusters in order to establish the 'affinity' of the query with clusters. This 'comparison' can be carried out, for instance, by computing a similarity $S_j$ between the above query vector $\mathbf{q}$ and the class centroid $\mathbf{r}_j$, for instance as follows (dot product):

$$S_j = \sum_{s=1}^{t} a_{js} q_s, \quad j = 1, \ldots, m \tag{3.103}$$

and a cluster $k_j$ is returned whenever the similarity exceeds some threshold $T$:

$$S_j \geq T \tag{3.104}$$

In this way several clusters can be retrieved in return to a query. In addition, even those documents may be furnished which are reached through $b_{ip}$ and $b_p$, documents found upon terms cluster (part of a thesaurus).

### Notes
In the cluster paradigm of *IR*, it is typically assumed (so called *cluster hypothesis*) that (1) topically related documents should form a cluster, and (2) the documents of a cluster should be relevant to the same query. Recent research shows [see e.g. Shaw, Burgiu and Howell, 1997] that the cluster hypothesis is debatable, at least where it is based on fixed clustering. Instead, adaptive clustering should be used ($I^2R$'s idea of flexible interconnections).

## EXAMPLE 3.6
Consider Example 3.2. Let $w_{12} = 0.5$, $w_{13} = 0.5$, $w_{23} = 0.8$ be the elements of the similarity matrix (recall: it is symmetric). Let the threshold $T = 0.7$. Then there are two clusters: $k_1 = \{o_1\}$ and $k_2 = \{o_2, o_3\}$. Assume the terms form a cluster each. Let the corresponding document vectors be as follows (occurrences): $v_1 = (1, 1, 0, 0, 1)$, $v_2 = (0, 1, 1, 0, 1)$, $v_3 = (0, 1, 0, 1, 0)$. The cluster representative $r_2$ for $k_2$ is as follows: $r_2 = (0, 1, 0.5, 0.5, 0.5)$. For retrieval purposes, a similarity is computed between the query vector $v_q = (0, 1, 1, 0, 0)$ and the two cluster representatives $r_1 = (1, 1, 0, 0, 1)$ and $r_2$.

## 3.2 Fuzzy Model

The Fuzzy model of *IR* is a 'softening' of the classical Boolean *IR*. In this model, a document is represented as a fuzzy set of terms, i.e., a set of pairs (term, membership_function_$\mu$(term)).

Typically, queries are boolean queries. (Query terms can also have user weights reflecting their importance. This is referred to as weighted Boolean model, and will be dealt with in Chapter 4.)

Retrieval is based on evaluating functions (measures) to retrieve documents in response to each query term, and on fuzzy operations to generate a final answer, as a whole.

Let

$$D = \{D_1, ..., D_i, ..., D_n\} \tag{3.105}$$

be a set of elements called *documents*, and

$$F_1, ..., F_j, ..., F_m \tag{3.106}$$

be fuzzy sets representing *subject areas*.

In this case, it is possible to consider the following membership function:

$$0 \le f_j(D_i) \le 1 \qquad\qquad\qquad (3.107)$$

representing the degree to which document $D_i$ pertains to fuzzy set $F_j$. For instance, $F_j$ could be the concept class represented by the index term $t_j$, and the membership function represents then a weight $w_{ij}$ of term $t_j$ in document $D_i$.

Each document $D_i$ is thus represented by a vector:

$$(f_1(D_i), ..., f_j(D_i), ..., f_m(D_i)) \qquad\qquad\qquad (3.108)$$

or, equivalently, by a vector $\mathbf{v}_i$ of weights $w_{ij}$ as follows:

$$\mathbf{v}_i = (w_{i1}, ..., w_{ij}, ..., w_{im}) \qquad\qquad\qquad (3.109)$$

Then the weights $w_{ij}$ may be interpreted as the extent to which document $D_i$ pertains to term $t_j$. In other words, $w_{ij}$ may be seen as a membership function (in fuzzy logic terminology) of documents $D_i$ to the subject area (or category) represented by the term $t_j$. Thus we also have the following fuzzy sets:

$$t'_j = \{(D_i, w_{ij}) | i = 1, ..., n\}, \quad j = 1, ..., m \qquad\qquad\qquad (3.110)$$

Thus, each term $t_j$ may be seen as a representative term of a subject class, which is conceived as a fuzzy set in the domain of documents $D_i$.

Let $Q$ denote a Boolean query (in a conjunctive normal form, for example):

$$Q = \bigwedge_s (\bigvee_k \theta_j), \theta_j \in \{t_j, \neg t_j\} \qquad\qquad\qquad (3.111)$$

(see *BIR*, too). Since the query $Q$ can always be transformed into a disjunctive or conjunctive normal form, it is sufficient to consider the following basic forms:

- single–term query: $A$
- OR–ed terms query: $A$ OR $B$
- AND–ed terms query: $A$ AND $B$
- negated term query: NOT $A$

Any other Boolean query is a combination of these basic forms. Query terms are matched against index terms $t_j$. In the case of exact matches, retrieval can take place, but in terms of fuzzy logic operators as follows, one by one:

**Single–term query**

In the case of a single term, those documents from the fuzzy set

$$A' = \{(D_i, w_{iA})\} \qquad (3.112)$$

are returned for which $w_{iA}$ exceeds a threshold (which may even be null, of course).

**AND–ed query**

In the case of AND–ed terms, we proceed as follows:

a) Firstly the corresponding fuzzy sets $A'$ and $B'$ are obtained (single–term query).
b) The corresponding fuzzy set for intersection is:

$$A' \text{ AND } B' = \min \{(D_i, w_{iA}), (D_i, w_{iB})\} \qquad (3.113)$$

c) Those documents are returned which occur in the fuzzy set above (a threshold can be used, too).

**OR–ed query**

In the case of OR–ed terms, we proceed as follows:

a) Firstly the corresponding fuzzy sets $A'$ and $B'$ are obtained (single–term query).

b) The corresponding fuzzy set for union is:

$$A' \text{ OR } B' = \max \{(D_i, w_{iA}), (D_i, w_{iB})\} \qquad (3.114)$$

c) Those documents are returned which occur in the fuzzy set above (a threshold can be used, too).

**NOT–ed query**

In the case of a negated term, we proceed as follows:

a) Firstly the corresponding fuzzy set $A'$ is obtained (single–term query);
b) The corresponding fuzzy set for negation is obtained:

$$\text{NOT } A' = \{(D_i, 1 - w_{iA})\} \tag{3.115}$$

c) Those fuzzy sets (concretely: columns in the matrix) are located which are the closest (within an error) to the fuzzy set $A'$.
d) Those documents are returned which occur in the fuzzy sets thus located, or for which the corresponding weights exceed a threshold.

It can be easily checked that the fuzzy model of $IR$ contains, as a special case, the classical Boolean model.

EXAMPLE 3.7
Consider Example 3.2. The fuzzy sets are (see 3.110): $t_1 = \{(D_1, 1/3), (D_2, 0), (D_3, 0)\}$, $t_2 = \{(D_1, 1/3), (D_2, 1/3), (D_3, 1/2)\}$, $t_3 = \{(D_1, 0), (D_2, 1/3), (D_3, 0)\}$, $t_4 = \{(D_1, 1/3), (D_2, 1/3), (D_3, 0)\}$. Let the query be $Q = t_2$ AND $t_3$. Then the answer will be $D_2$.

## 3.3    Latent Semantic Indexing Model

From a purely formal mathematical point of view, *Latent Semantic Indexing* (*LSI*) represents a vector *IR* model. However, *LSI* is different from the classical *VIR* in the way in which the weight vectors, representing documents and queries for retrieval purposes, are generated (or, in other words, in the way in which indexing is performed).

In principle, *LSI* derives 'artificial concepts' representing common meaning components of documents, these being represented by weight vectors indicating a level of association between documents and these concepts. This representation is computationally economical, i.e., the dimension of vectors is much less than the number of terms (much less than, e.g., in the classical *VIR*), and better captures common meaning in documents.

Let

$$D = \{D_1, ..., D_i, ..., D_n\} \tag{3.116}$$

be a set of elements called *documents*, and

$$T = \{t_1, ..., t_j, ..., t_m\} \tag{3.117}$$

a set of elememnts called *terms*. Let

$$W = (w_{ij})_{n,m} \qquad\qquad (3.118)$$

be a *weights matrix* where $w_{ij}$ denotes a weight of term $t_j$ in document $D_i$, e.g., number of occurrence, or some other weight. Let the rank of $W$ be $r$, i.e., rank($W$) = $r$, and the singular value decomposition of $W$ be

$$W = USV^T \qquad\qquad (3.119)$$

where $U$ corresponds to term vectors, $S$ to singular values and $V$ to document vectors. Thus the singular value decomposition of $W$ represents a breakdown of the original relationships ($W$) between documents and terms into factor values (linearly independent vectors). In other words, a set of uncorrelated indexing variables (artificial concepts) are obtained corresponding to a factor value $k = 2, 3, ...$ ($k$ is the number of columns in $U$, of rows in $S$ and $V$). Thus

$$W_k = U_k S_k V_k^T \qquad\qquad (3.120)$$

is an approximation of the original matrix $W$ with the weights of 'artificial concepts'. Of course, if $k = n$ then $W_k = W$. The matrix $W_k$ is used for retrieval purposes. Similarly, for retrieval purposes, a query $q$ is represented as a weights vector as follows:

$$q_k = q^T U_k S_k^{-1} \qquad\qquad (3.121)$$

where $q^T$ is the vector of frequencies of terms in $q$, whereas $U_k$ and $S_k$ are the respective weights.

Retrieval is performed by computing the value of a similarity (association) measure (e.g., cosine, dot product) between the vectors $q_k$ and $W_k$. A threshold value or rank order can also be used.

EXAMPLE 3.8
Consider the following documents:

$D_1$ = *Bayes' Principle: The principle that in estimating a parameter one should initially assume that each possible value has equal probability (a uniform prior distribution).*

$D_2$ = *Bayesian Conditionalisation: This is a mathematical procedure with which we revise a probability function after receiving new evidence. Let us say that we have probability function P(.) and that through observation I come to learn that E. If we obey this rule our new probability function Q(.) should be such that for all X Q(X)=P(X|E) we are then said to have 'conditionalised on E'.*

$D_3$ = *Bayesian Decision Theory: A mathematical theory of decision making which presumes utility and probability functions, and according to which the act to be chosen is the Bayes act, i.e., the one with highest Subjective Expected Utility. If one had unlimited time and calculating power with which to make every decision, this procedure would be the best way to make any decision.*

$D_4$ = *Bayesian Epistemology: A philosophical theory which holds that the epistemic status of a proposition (i.e. how well proven or well established it is) is best measured by a probability and that the proper way to revise this probability is given by Bayesian conditionalisation or similar procedures. A Bayesian epistemologist would use probability to define, and explore the relationship between, concepts such as epistemic status, support or explanatory power.*

Let the terms be as follows:

$t_1$ = *Bayes' Principle*
$t_2$ = *probability*
$t_3$ = *Bayesian Conditionalisation*
$t_4$ = *decison making*

The matrix $W$ is as follows:

$$W = \begin{bmatrix} 1 & 0 & 0 & 0 \\ 1 & 3 & 1 & 3 \\ 0 & 1 & 0 & 1 \\ 0 & 0 & 1 & 0 \end{bmatrix}$$

The singular value decomposition of $W$ is $W = USV^T$:

$$U = \begin{bmatrix} -0.046 & 0.644 & -0.707 & -0.289 \\ -0.955 & 0.062 & 0 & 0.289 \\ -0.288 & -0.409 & 1.591 \cdot 10^{-15} & -0.866 \\ -0.046 & 0.644 & 0.707 & -0.289 \end{bmatrix}$$

$$S = \begin{bmatrix} 4.68 & 0 & 0 & 0 \\ 0 & 1.047 & 0 & 0 \\ 0 & 0 & 1 & 0 \\ 0 & 0 & 0 & 0 \end{bmatrix}$$

$$V = \begin{bmatrix} -0.214 & 0.674 & -0.707 & 0 \\ -0.674 & -0.214 & 0 & -0.707 \\ -0.214 & 0.674 & 0.707 & 0 \\ -0.674 & -0.214 & 0 & 0.707 \end{bmatrix}$$

Taking factor value $k = 2$, $W_k$ is as follows:

$$W_k = \begin{bmatrix} -0.408 & -0.288 & 0.151 & -0.477 \\ 0.913 & -3.028 & 3.162 & -0.046 \\ 0.577 & -0.817 & 0.953 & 0.303 \\ -0.408 & -0.288 & 0.151 & -0.477 \end{bmatrix}$$

Consider now the query $q$ consisting of two terms as follows:

$q = probability, \ decison \ making$

The corresponding query vector is:

$$[0 \ 1 \ 1 \ 0] \cdot U_k^T \cdot \text{diag}(S_k)^{-1} = [0.013 \ -0.39]$$

which is to be compared — in terms of a similarity measure — with $U_k$.

## 3.4     Artificial Intelligence in Information Retrieval

### 3.4.1     Artificial Neural Networks Model

A 3–layer network $(Q, T, D)$ — to enhance the Probabilistic *IR* — is considered, where $Q$ and $D$ denote the layer of queries $Q_k$ and of documents $D_i$, respectively. Queries and documents are regarded as neurons of the same category, and can act either as input or output units. $T$ denotes the layer of index terms $t_j$ which are connected bi–directionally with documents and queries. (Figure 3.9)

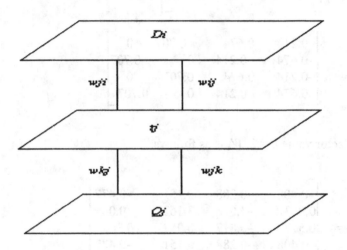

*Figure 3.9.* A 3–layer ANN for IR

Intra–layer connections are disallowed. Both the output and   transfer functions are taken as the identity function. The net is considered to be initially blank. As each document neuron is created and fed into the network, associated index term neurons are also created. From the document neuron $D_i$ to the index (new or existing) term neuron $t_j$ a new connection is created and its strengths $w_{ij}$ is set to:

$$w_{ij} = \frac{f_{ij}}{L_i} \qquad\qquad\qquad (3.122)$$

where $f_{ij}$ denotes the number of occurrences of term $t_j$ in document $D_i$, and $L_i$ denotes the length (e.g. total number of index terms) of that document.

From a term neuron $t_j$ to a document neuron $D_i$

a) A new connection is created if that term is new, and its strengths $w_{ji}$ is set to (a special form of inverse document frequency; it also can be derived using Bayes' Formula from the Probabilistic *IR*; see e.g. [Kwok, 1989]):

$$w_{ij} = \log \frac{p}{1-p} + \log \frac{1 - s_{ji}}{s_{ji}} \qquad (3.123)$$

where $p$ is a small positive constant (the same for all terms), and $s_{ji}$ is the proportion $F_j/N_w$ where $F_k$ represents the document collection number of occurrences of the $j$th term, and $N_w$ is the total number of terms in the document collection.

b) If term $t_j$ already exists, then all connections emanating from it to documents have to be adjusted, too.

Queries $Q_k$ are treated as if they were documents. Retrieval is viewed in the following way:

a) Let us consider a query neuron $Q_k$, and clamp its activity to 1.
b) Any term neuron $t_j$ affected by this query will receive as input:

$$1 \cdot w_{kj} \qquad (3.124)$$

and outputs the same value (identity transfer and output function);
c) From each of the terms affected by the query some documents $D_i$ (those having terms in common with the query) are affected receiving input $I_i$:

$$\sum w_{ji} w_{kj} \qquad (3.125)$$

(summation over $j$) and output the same value $I_i$ (again, identity transfer and output function);
d) The documents so affected turn back the activation to term neurons which receive input $I_j$:

$$\sum I_i w_{ij} \qquad (3.126)$$

(summation over $i$) and which output the same value $I_j$ (again, identity transfer and output function) to query neurons;
e) The output values obtained at point d) can then be used for ranking purposes, or for relevance evaluation through a man–machine dialogue, or can be compared with a threshold.

EXAMPLE 3.9
The following simple example ($p = 1/2$) illustrates weights calculation and activation spreading for ranking purposes. Let the first document be ($i = 1$):

$D_1$ = Bayes' Principle: The principle that in estimating a parameter one should initially assume that each possible value has equal probability (a uniform prior distribution).

having the following term, $j = 2$: $t_1$ = Bayes' Principle, $t_2$ = probability. The connections are as follows:

$w_{ij}, i = 1, j = 1, 2$: $w_{11} = 1/2, w_{12} = 1/2,$
$w_{ji}, j = 1, 2, i = 1$: $w_{11} = 0, w_{21} = 0.$

Let the query be $q_k$ = Bayes' Principle, for example ($k = 1$). The corresponding wights are:

$w_{kj}, k = 1, j = 1, 2$: $w_{11} = 1, w_{12} = 0,$
$w_{jk}, j = 1, 2, k = 1$: $w_{11} = -\infty, w_{21} = 0.$

Activation spreading (retrieval) is then straightforward: from $q_k$ an activation with value $1 \cdot w_{kj} = 1 \cdot w_{11} = 1$ is started, it is spread over to $D_1$ having value $w_{kj}w_{ji} = w_{11}w_{11} = 1$. From $D_1$, the activation is turned back to the term it has started from, having values 1/2. Were there several documents, a ranking could be made at this point.

### 3.4.2    Genetic Algorithms Model

Genetic Algorithms ($GA$) are concerned with using genetic concepts and techniques to problem solving.

#### 3.4.2.1    Basic Concepts

Genetic Algorithms ($GA$) are special algorithms making use of principles of evolution and heredity.
    A population

$$P(t) = x_1 x_2 ... x_i ... x_n$$

of *individuals* (chromosomes) $x_i$, $i = 1, ..., n$, is maintained at each iteration (time, step) $t$.

Every individual $x_i$ is a sequence of elementary units called *genes*, and represents a potential solution to the problem to be solved, and evaluated to obtain a measure of *fitness*.

At a next iteration $t + 1$, a new population is *selected* (or formed) by selecting the fitter individuals.

Some individuals undergo transformations such as *mutation* (*m*) or *crossover* (*c*):

- Mutation means a small change in a single individual and thus yields a new individual.
- Crossover leads to new individuals by combining parts of two or more individuals.

For example, the crossover between the individuals

    *abcd*

and

    *xyuv*

after the second gene produces the offsprings

    *abuv*

and

    *xycd.*

In a *GA*, the crossover and mutation operations are control parameters and need to be carefully defined to obtain a solution. After a number of iterations, the *GA* converges and the best (fittest) individual, hopefully, represents the optimal solution to the problem at hand.

### 3.4.2.2 Information Retrieval

When applying *GA* to *IR*, the following correspondances are made:

| gene | term (keyword, concept) |
|------|------------------------|
| individual | document |
| population | set of documents |

Let the total number of genes in each chromosome be $g$. In practice a gene is a bit with the value 1 indicating the presence of the term or 0 for the absence. An initial population, having size $s$, is given by, e.g., a relevance feedback from the user. The problem to be solved using *GA* is to find an optimal population, i.e., genes that best describes the initial population and can then be used by an *IR* system. The user–*GA*–*IR* system interaction can be continued until, e.g., the user decides to stop. Thus the *GA* based model of *IR* can be used, for example, to complement a classical one.

The following steps implement a typical *GA* for *IR*:

*Step* 1.
A fitness measure is based on a similarituy, e.g., Jaccard's coefficient $J$ which is (in a Set Theoretic form):

$$J = \frac{|A \cap B|}{|A \cup B|} \qquad (3.127)$$

where $A$ and $B$ are sets and $|\,.\,|$ denotes cardinality. The individuals that should undergo transformations are selected randomly.

*Step* 2.
Now, crossover can be applied to the new (selected) individuals. Given a probability $p_c$ of crossover, $s \times p_c$ is the number of individuals that should undergo crossover. For each chromosome, a random number $r \in (0, 1)$ is generated, and, if $r < p_c$, the respective chromosome is selected for crossover. Then the selected pairs of chromosomes are mated randomly for crossover. A random number $q \in (1, g - 1)$ indicates the position of the crossover point, and the respective chromosomes exchange genes accordingly.

*Step* 3.
The other genetic operation, mutation, can now also be performed. Given a probability $p_m$ of mutation, $p_m \times g \times s$ is the number of mutated genes. Every gene in the whole population has an equal chance to undergo mutation. For each chromosome in the crossovered population, and for each gene within every chromosome, a random number $r \in (0, 1)$ is generated, and, if $r < p_m$,

the respective gene is mutated, i.e., changed from 1 to 0 or vice versa, accordingly.

Steps 1, 2 and 3 are repeated cyclical until convergence is reached (i.e., no improvement in the overall fitness of the population) or a predetermined number of generations is reached.

EXAMPLE 3.10
Let the initial population $D$ (given by relevance feedback or other method) be

$$D = \{D_1, D_2, D_3\}$$

where

$D_1$ = *Bayes' Principle: The principle that in estimating a parameter one should initially assume that each possible value has equal probability (a uniform prior distribution).*

$D_2$ = *Bayesian Decision Theory: A mathematical theory of decision making which presumes utility and probability functions, and according to which the act to be chosen is the Bayes act, i.e. the one with highest Subjective Expected Utility. If one had unlimited time and calculating power with which to make every decision, this procedure would be the best way to make any decision.*

$D_3$ = *Bayesian Epistemology: A philosophical theory which holds that the epistemic status of a proposition (i.e., how well proven or well established it is) is best measured by a probability and that the proper way to revise this probability is given by Bayesian conditionalisation or similar procedures. A Bayesian epistemologist would use probability to define and explore the relationship between concepts such as epistemic status, support or explanatory power.*

Which are the terms that best best describe the information need expressed by the user in these documents? Let the total set $T$ of terms be

$$T = \{\text{Bayes' Principle, probability, decision making}\}, \ g = 3.$$

The initial genetic patterns of chromosomes in population is as follows:

chromosome $C_1 = 1\ 1\ 0$

chromosome $C_2$ = 0 1 1
chromosome $C_3$ = 0 1 0

An average fitness is calculated for each document, as follows: calculate Jaccard's coefficient between any pair $D_i$ and $D_j$, $j$ =1, 2, 3, first, and for every $i$ = 1, 2, 3, and then take the average. Thus, the following average fitnesses are obtained:

chromosome $C_1$ = 1 1 0; average fitness = 0.6
chromosome $C_2$ = 0 1 1; average fitness = 0.6
chromosome $C_3$ = 0 1 0; average fitness = 0.66
                         Average fitness for the population: 0.62

Let chromosomes $C_1$ and $C_2$ be the ones to undergo crossover after the second gene. The new population is as follows:

chromosome $C_1$ = 1 1 1; average fitness = 0.55
chromosome $C_2$ = 0 1 0; average fitness = 0.77
chromosome $C_3$ = 0 1 0; average fitness = 0.77
                         Average fitness for the population: 0.69

It can be seen that the overall fitness of the population has increased (by 0.07). If one decided to stop here and perform a search, using an *IR* system, the term "probability" (highest average fitness) could be used to perform the search. After that, using the documents thus retrieved, the *GA*–procedure can be repeated.

*GA*–based *IR* performs acceptably well especially when the initial set of population is fuzzy. This is an important feature for implementing practical searches in large databases which is a difficult problem.

### 3.4.3    Knowledge Bases, Natural Language Processing

*Knowledge Bases* (*KB*) and *Natural Language Processing* (*NLP*) can be used to enhance an *IR* system (e.g., vector *IR*) by providing methods for e.g., *query expansion*, i.e., for providing methods to automatically generate queries that are equivalent (e.g., using synonyms) or similar (e.g., using related terms) to the user's original query. Thus retrieval performance is expected to increase.

### 3.4.3.1   Natural Language Processing

In many cases documents in *IR* (journal articles, newspaper articles, TV news texts, etc.) are natural language texts. They have a language structure characterised by dependency between words and terms yielding meanings of texts. In a way classical *IR* — and especially classical indexing and weighting of terms — was a first attempt to apply *Natural Language Processing* (*NLP*) techniques to *IR*. *NLP* techniques are inherently difficult ones, and they are usually applied to enhance *IR* models by trying to discover structures of texts which allow computation and preserve meaning.

Some of the most typically used are as follows.

*Stemming*. Stemming in *IR* means to reduce words to their roots (see, e.g., the well known Porter Algorithm). In stemming, a common basic structure is discovered and several words are reduced to this structure, which thus can replace them in implemented retrieval algorithms. For example, the words *information*, *informational* can be reduced to their stem *inform*. Thus, the stem can be used as an indexing as well as a search term.

*Frames, patterns, samples*. These are — discovered or pre–defined — fixed structures filled with appropriate text parts. Thus, retrieval is based on comparing (not necessarily string matching) document frames with query frames. Such techniques are used, for example, by Croft and Lewis [1987], Croft and Thompson [1987], or in Internet retrieval [e.g. Guan and Wong, 1999]. In some cases, e.g., Croft et al. above, *NLP* techniques are applied to a homogeneous domain, namely, science and technology, and a knowledge base is kept with language structures.

*Networks* (conceptual graphs, hierarchy). They represent texts as a network of interconnected words, terms or text surrogates along with relationships between them. Such techniques are used, for example, in thesauri and Internet retrieval [see, e.g., Cohen and Fan, 1999; Guan and Wong, 1999].

### 3.4.3.2   Semantic Network

From a mathematical point of view a semantic network is a labelled weighted directed multigraph. The vertices (nodes) of the graph represent, for example, words or concepts (e.g., in a stemmed format). Frequencies of words can also be included in the nodes. The edges (links) signify relationships between words such as synonyms, broader, narrower or related term (in meaning), IS_A connection, etc. (e.g., thesaurus). In particular, the model of the semantic network can be a tree, representing a hierarchy of word classes.

### 3.4.3.3    Knowledge Base

A *Knowledge Base* (*KB*) is a representation and collection of knowledge usually specific to an application area. One way in which to represent knowledge is by using semantic networks (other knowledge representation forms are *rules, frames*, etc.). An example for such a *KB* is a *thesaurus* (for example, WordNet, see below).

### 3.4.3.4    Information Retrieval

In what follows, three typical applications are briefly described.

### 3.4.3.4.1    Query Expansion

Given, for example, a vector *IR* model. In addition to this, let $\beta$ be an associated *KB* using a semantic network of terms (e.g., WordNet [Miller, 1990], or a domain–specific *KB* constructed automatically, [e.g., Frakes and Baeza–Yates, 1992; Bodner and Fei Song, 1996]). The use of $\beta$ is as follows.

Given the user's original query $Q$ (consisting of terms and represented as a vector of weights).

The semantic network can be used to find broader terms (e.g., following links up the hierarchy towards the root). Thus $Q$ can be expanded to broader terms to allow matching with a greater range of documents. It is expected that more relevant documents can thus be retrieved (higher *recall*).

The semantic network can be used to find narrower terms (e.g., following links down the hierarchy towards leaves). Thus $Q$ can be expanded to narrower terms to allow matching with a smaller range of documents. It is expected that the proportion of the retrieved and relevant documents becomes larger (higher *precision*).

The semantic network can be used to find related terms. Thus $Q$ is expanded to allow matching of as many relevant documents as possible.

Matching is done in all cases using a standard similarity measure.

### News

Watanabe et al. [1999] report on using *NLP* to enhancing a news based *IR* system. First, text is extracted from TV news report $x$. Then using morphological analysers, nouns $k$, $k = 1, ..., n$, are extracted from this text. The TV news is assigned the (corresponding) same day's newspaper article $y$

(collected from the World Wide Web WWW, because it is in an electronic format and has tags which help identify text parts). A score $S(x, y)$ is calculated between the TV news and article as follows:

$$S(x, y) = \sum_{k=1}^{n} \sum_{i=1}^{4} \sum_{j=1}^{2} w_{ij} \, f_{TV,ik} \, f_{newspaper,jk} \, l_k \qquad (3.128)$$

where
- $w_{ij}$ is the weight of word $k$ in location $i$ (headline, picture caption, first paragraph, the rest) of newspaper and location $j$ (title, the rest) of TV news. For example, if the word $k$ appears in the title of the TV news and in a picture paragraph in the newspaper, then $w_{ij} = 4$.
- $f_{TV,ik}$ and $f_{newspaper,jk}$ are the number of occurrences of word $k$ in location $i$ (headline, picture caption, first paragraph, the rest) of newspaper and location $j$ (title, the rest) of TV news, respectively.
- $l_k$ is the length of word $k$.

If $S(x, y)$ exceeds a threshold then the TV news and the newspaper article are 'aligned' (connected) to each other.

Retrieval (full text) is performed as follows.

- The user enters a query word;
- Firstly, the titles and dates of newspaper articles which contain the query words are shown (Boolean *IR*);
- Secondly, all aligned TV news reports are also shown.

### 3.4.3.4.2 Image Retrieval

Favela et al. [1999] report on an application that integrates image content and text.

Images are identified on the World Wide Web (WWW) along with their HTML text.

An image has an associated text if:

- it has an associated caption HTML tag;
- it has a text to its side;
- it is the only content of that page.

From this text stop words are excluded first. For the remaining words the inverse document frequency formula is used to compute their weights, and then the highest ten words are kept as image descriptors.

Further, to each image a special transformation is applied (a so called *wavelet transform*, see, e.g., Stollnitz et al. 1995), which yields 256×256 coefficients, out of which the highest 80 are kept as identifying the image.

Thus a Web digital library was created and stored in a database, containing plants and animals.

The user enters his/her query in the form of keywords or image or both, which are first transformed into vectors of weights (see above). A dot product similarity function is used to retrieve texts or images. If the query contains both image and text, the similarity is a weighted sum of the two dot products. (These weights are provided by the user, and reflect his/her preference for text and image.)

### 3.4.3.5    Thesauri Federation

With the advent of, e.g., Internet, WWW, and European Union, the importance of *cross–language retrieval* has increased, i.e., a retrieval situation where the languages of the documents and queries can differ. One technique used in cross–language retrieval is the use of a *thesaurus federation*: several thesauri (e.g., in different areas and langauges) are loosely cupled (as a 'meta–graph') so as to allow inter–thesauri navigation.

## 3.5    Bibliographical Remarks

The roots of the fuzzy model can be traced back to the 1970s [see e.g. Negoita and Flondor, 1976; Radecki, 1979; Waller and Kraft, 1979; Bookstein, 1980; Salton and McGill, 1983]. Properties and applications are studied in e.g [Kraft, 1985; Buell, 1985; Dubois et al., 1988; Pao, 1989; Myamoto, 1990, 1998; Kraft and Buell, 1992; Bordogna and Pasi, 1993; Kraft and Bordogna, 1998; Kraft and Monk, 1998; Pasi, 1999].

Kraft, Bordogna and Pasi [1998] offer an excellent overview of fuzzy techniques used in all major areas of *IR*.

*ANN* techniques are used in *IR* in, e.g., [Kwok, 1989; Chen, 1995]. Also, the Hopfield *ANN* is applied to *IR*.

The use of Genetic Algorithms for *IR* is nicely presented in, e.g., [Chen, 1995].

Knowledge–based techniques (expert systems, semantic networks) in *IR* are investigated and applied in e.g. [Smeaton and van Rijsbergen, 1983;

Forsyth and Rada, 1986; Brooks, 1987; Frakes and Baeza–Yates, 1992; Crouch and Young, 1992; Croft, 1993; Qui and Frei, 1993; Vorhees, 1994; Liu, 1995].

Wordnet [Miller, 1990] is an on–line dictionary based on a semantic network; it is a general–world knowledge base.

WebKB [Martin and Eklund, 1999] is accessible over the WWW, and is a knowledge–based (using conceptual graphs) retrieval engine. It also uses lexical techniques to generate and retrieve knowledge in WWW documents.

*LSI* is presented and treated in, e.g., [Deerwester et al., 1990; Berry and Dumais, 1994; Baeza–Yates and Ribeiro–Neto, 1999].

Further properties can be found in e.g. [Ford, 1991; Croft, 1993; Chen, 1994, 1995; Strzalkowski, 1995; Kang and Choi, 1997; Croft and Ponte, 1998; Chen, Mikulic and Kraft, 1998; IEEE Intelligent Systems, Special Issue on Intelligent *IR*, Vol 14. No. 5, 1999].

## 4.    TRADITIONAL TREE STRUCTURE OF INFORMATION RETRIEVAL MODELS

Figure 3.10 shows a structural schematic of the *IR* models (tree). We will see, after the mathematical foundation in Chapter 4, that this schematic will become more consistent formally.

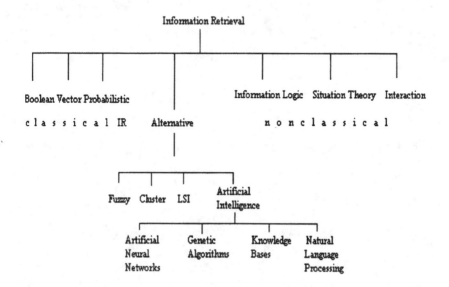

*Figure 3.10* Schematic of *IR* models.

# Chapter 4

# MATHEMATICAL THEORY OF INFORMATION RETRIEVAL

There exist different *IR* classes of models: classical, nonclassical, alternative (Chapter 3). Because the alternative models are based on classical models (of which they are enhancements) using special techniques from different fields (such as Fuzzy Sets, Artificial Intelligence, etc.), the classical models bear fundamental importance. Thus, a mathematical foundation and theory for the classical models serves, partly, as a theoretical basis for the alternative models, too. Moreover, a formal hierarchical link with the nonclassical model of $I^2R$ will also be shown (namely: the classical models can be formally conceived as a special case of $I^2R$), and this may be viewed as an expression of an internal structural consistency and dynamics in the formal development of *IR* models, as a whole.

Formal links to modern mathematical disciplines (Metric Spaces, Topology, Vector Spaces, Matroid Theory, Recursion Theory, Complexity Theory, Decision Theory) are consistently created with the result new properties and insights of *IR* can be revealed .

The approach adopted in this chapter is somewhat axiomatic — without over–emphasising the advantages or disadvantages of such a method [see, e.g., Tarski, 1956; Lakatos, 1976].

## 1. INFORMATION RETRIEVAL FRAME

A widely accepted — not mathematical — definition of Information Retrieval (*IR*) is the following: *IR* is concerned with the organisation, storage, retrieval and evaluation of information that is likely to be relevant to users' information needs. Although it seems impossible to give a mathematical definition for *IR*, mainly because of the subjective nature of relevance and the difficulties in defining the concept of information (see, e.g., [van Rijsbergen, 1979; Mizzaro, 1994]), a formal and generic description has taken shape, namely: *IR* model, which serves as a starting point for an algorithmic view on *IR* models [see e.g. Baeza–Yates and Ribeiro–Neto, 1999].

Information can be looked for in titles, abstracts, articles, images, pieces of sound, hypermedia. These can be generically referred to as *documents*. For the retrieval — methods, techniques, procedures, algorithms — to work, different *representations* of documents are used (to allow computation with them), such as, for example, sets of keywords, vectors of weights, *n*–grams, language patterns, graphs, etc.. Because, in applications, there always is a correspondence between documents and their representations, they both can be treated, in principle, as *objects*, in general. This will not cause any inconvenience or confusion, since we are only concerned with purely formal aspects, and it will always be made clear whether we mean documents or representations. For the same reason (formal aspects), as well as because it is the case in practice, the same conceptual space will be used both for objects and queries — these latters being conceived, for retrieval purposes, as objects having the same type as documents. We should note, however, that should one want to go beyond purely formal aspects, such as, e.g., the specific meaning of a query, taking the same conceptual space would invalidate some of the basic *IR* concepts, i.e., would yield meaningless similarity, term independence and linearity [see, e.g., Bollmann–Sdorra and Raghavan, 1993].

Thus a very abstract and broad generic concept modelling *IR* can be defined as a pair consisting of objects and a function (or mapping) that associates ('retrieves') some objects to an object (representing a query).

Let

$$D = \{d_1, ..., d_j, ..., d_M\}, M \geq 2 \qquad\qquad (4.1)$$

be a non–empty finite set of elements called *objects*.

Notes

The case $M = 1$ can, of course, be mathematically considered, but it would be trivial. Typically the objects are representations.

Let $\Re$ be a mapping, called *retrieval*, from $D$ into its power set $\wp(D)$, i.e.,

$$\Re: D \to \wp(D) \tag{4.2}$$

For example, $\Re$ can be a similarity association measure, an inference process, etc. (see Chapter 3). By associating the set $D$ of objects and retrieval $\Re$, the following structure is defined:

*DEFINITION 4.1*
A 2–tuple $\Phi = \langle D, \Re \rangle$ is called an *Information Retrieval Frame* (*IRF*).

Definition 4.1 is a very general one: it does not say anything about possible specific forms of retrieval $\Re$ and objects $D$. This is a point from which different specific *IR* models can be obtained, by particularising $D$ and $\Re$.

Usually a specific *IR* model is tested (using a *test collection*, such as, e.g., ADI, TIME, MEDLINE, CRANFIELD, TREC, etc.), and before being applied to a real collection. In both cases, the same retrieval method is used, although the sizes of objects and collections can be very different (the size of the test collection is typically much smaller). This makes it possible to define a relation between frames as follows (one's retrieval is a restriction, to a smaller set, of the other's):

*DEFINITION 4.2*
Let $\Phi = \langle D, \Re \rangle$ be an *IRF*. Two *IRF*s, $\Phi_1 = \langle D_1 \subseteq D, \Re_1 \rangle$ and $\Phi_2 = \langle D_2 \subseteq D, \Re_2 \rangle$, are $\Re$–*equivalent*, notation $\Phi_1 \equiv \Phi_2$, if $\Re_1 = \Re|D_1$ and $\Re_2 = \Re|D_2$.

Mathematically the assumption (or hope) that the same *IR* model will exhibit a similar behaviour when applied to collections of different sizes (for example, changing from a test collection to a real one) is based on (or expressed in) the following:

**THEOREM 4.1**
Relation $\equiv$ is an equivalence relation.

**Proof**. a) $\equiv$ is reflexive, $\Phi_1 \equiv \Phi_1$: $\Re_1 = \Re_1 = \Re|D_1 \Rightarrow \Phi_1 \equiv \Phi_1$. b) $\equiv$ is symmetric, $\Phi_1 \equiv \Phi_2 \Rightarrow \Phi_2 \equiv \Phi_1$: $\Phi_1 \equiv \Phi_2 \Leftrightarrow \Re_1 = \Re|D_1 \wedge \Re_2 = \Re|D_2 \Rightarrow \Phi_2 \equiv \Phi_1$. c) $\equiv$ is transitive, $\Phi_1 \equiv \Phi_2 \wedge \Phi_2 \equiv \Phi_3 \Rightarrow \Phi_1 \equiv \Phi_3$: $\Phi_1 \equiv \Phi_2 \wedge \Phi_2 \equiv \Phi_3 \Leftrightarrow (\Re_1 = \Re|D_1 \wedge \Re_2 = \Re|D_2) \wedge (\Re_2 = \Re|D_2 \wedge \Re_3 = \Re|D_3) \Rightarrow \Re_1 = \Re|D_1 \wedge \Re_3 = \Re|D_3 \Rightarrow \Phi_1 \equiv \Phi_3$ ◆

An immediate and interesting consequence of Theorem 4.1 is the following:

**COROLLARY 4.1**
The equivalence relation $\equiv$ defines a partition on a set of *IRF*s. ◆

In other words, Corollary 4.1 means that all *IFRs* $\Phi_i = \langle D_i, \mathfrak{R}_i \rangle$ whose retrievals $\mathfrak{R}_i$ are restrictions of the same retrieval $\mathfrak{R}$ exhibit the same formal mathematical structure and behaviour (from a retrieval point of view) even for different sets of objects $D_i$. For example, all *IR* systems whose retrieval is based on the cosine measure, have the same mathematical structure. Similarly, for e.g. dot product, Dice's coefficient, etc.. Thus, for example, from a purely formal mathematical point of view, the SMART system and the *LSI* model bear the same mathematical structure (pseudo–metric induced topology) albeit they bear very important specific technical differences; for example, they use different techniques to generate index terms and weights vectors to represent objects.

Notes

Other formalisations of an *IR* system also exist. For example, Baeza–Yates and Ribeiro–Neto [1999] introduce the concept of an *IR model* which contains both the documents and their representations. Obviously, retrieval $\mathfrak{R}$ acts upon representations, and then, based on a well–defined correspondence between documents and their representations, the respective documents can be identified (and thus retrieved). Thus from a purely formal point of view however, it is not necessary to make such a disctinction. With the emerging Internet retrieval models, the correspondence between representations and documents can become *a posteriori* in that a retrieved document can emerge as a virtual rather than an existing entity in a collection [see, e.g., Watters, 1999]. Formally, the retrieved virtual document which emerges from or consists of existing parts of existing documents can be modelled by enlarging the set $D$ of documents with all subsets of all effectively existing documents or document parts (surrogates). From a purely formal mathematical point of view it seems that an *IRF* can be treated as an application of the theory of family of sets. Let $f: I \to \wp(D)$ define a family of sets, and $\mathfrak{R}: D \to \wp(D)$ define a retrieval. They can be made equivalent by taking $I = \{i|d_i\}$. Thus in principle an *IR* system is a family of sets directed both to the right and left. This property as well as many others, such as covering, separation, projection, hold too. They are in concordance with and reflected in the definition of *CIR* (see next below) and its topological properties.

# 2.    CLASSICAL INFORMATION RETRIEVAL

A unified definition (*CIR*) for the classical *IR* models (vector, probabilistic) is given as an abstract mathematical concept using that of an *IRF*.

## 2.1    Classical Information Retrieval

Let
1. $T = \{t_1, t_2, ..., t_k, ..., t_N\}$ be a finite set of elements called *identifiers*, $N \geq 1$. Identifiers can be, e.g., index terms, $n$–grams, image fragments, text surrogates, etc., and they serve to identify objects to be searched.

2. $O = \{o_1, o_2, ..., o_u, .., o_U\}$ be a finite set of elements called *objects, $U \geq 2$*;

3. $(D_j)_{j \in J = \{1,...,M\}}$ be a family of *object clusters, $D_j \in \wp(O), M \geq 2$*;

4. $D = \{õ_j | j \in J\}$ be a set of elements called *object documents* or *documents*, for short, where the normalised fuzzy set $õ_j = \{(t_k, \mu_{õ_j}(t_k)) | t_k \in T, \ k = 1, ..., N\}, j = 1,..., M, \mu_{õ_j}: T \to S \subseteq [0, 1] \subset \mathbf{R}$, is the *cluster representative* of object cluster $D_j$. For example, $O$ can consist of journal papers, each of which is a cluster itself, and each cluster's fuzzy representation is a document. In this case, if the fuzzy set is crisp, the document is identical to the classical binary vector representation. As another example, a cluster can be a collection of journal papers considered to be related (or similar) to each other, in which case the cluster representative (or document) is a fuzzy representation of, e.g., one of the papers or of a pseudo–paper (centroid, i.e., one which is not effectively one of those in the cluster but describes well the overall content of that cluster). Thus formally the traditional cluster *IR* model is formally contained here as a special case.

5. $A = \{ã_1, ..., ã_i, ..., ã_C\}$ be a finite set of elements called *criteria, $C \geq 1$*, where $ã_i = \{((q, õ_j), \mu_{ã_i}(q, õ_j)) | õ_j \in D, j = 1, .., M\}, i = 1, .., C$, is a normalised fuzzy relation, $\mu_{ã_i}: D \times D \to [0, 1] \subset \mathbf{R}, q \in D$ arbitrary fixed. Traditionally classical *IR* exhibits a dichotomic (two–pole) property, in that there are two distinct criteria: (i) which are either present or absent; and (ii) based on which retrieval is performed. But because, on the one hand, in recent years' research, this dichotomy has been softened (in, e.g., the Fuzzy *IR* model, see, e.g., [Kraft et al. 1998]), and, on the other hand, because a mathematical foundation makes a generalisation possible, it is assumed that, formally, there may be one, two, or more criteria (e.g. relevant, irrelevant, undecidable) each with different degrees (for example, half–relevant, more relevant than not, etc.). This is captured by conceiving criteria as fuzzy relations.

6. $a_{\alpha_i} = \{õ \in D | \mu_{ã_i}(q, õ) > \alpha_i\}, i = 1, ..., C$, be a strong $\alpha_i$–cut of criterion $ã_i$, $\alpha_i \geq 0, q \in D$ arbitrary fixed;

7. $\Re: D \to \wp(D)$ be a mapping called *retrieval*. Formally, retrieval means to associate a subset of documents to a query if they relate to each other — according to a selected criterion — strongly enough. This is captured by viewing the query as a document, and retrieval as being defined using $\alpha$–cuts [see, e.g., Negoita, 1973; Negoita and Flondor, 1976; Radecki,

1979; Waller and Kraft, 1979; Bookstein, 1980; Kraft, 1985; Buell, 1985; Dubois et al. 1988; Miyamoto, 1990; Bordogna and Pasi, 1993; Kraft et al. 1998].

A basic concept called *Classical Information Retrieval* (*CIR*), from which the classical *IR* models will be formally deduced, is defined as follows:

*DEFINITION 4.3* [DOMINICH 1999B]
An *IRF* $\Phi = \langle D, \Re \rangle$ with the following two properties P1 and P2:
  P1. $q = \delta \Rightarrow \mu_{a_i}(q, \delta) = 1$ (reflexivity), $\forall i, q, \delta$;
  P2. $\Re^i(q) = \{\delta | \mu_{a_i}(q, \delta) = \max_{k=1, ,C} \mu_{a_k}(q, \delta)\} \cap a_{\alpha_i}$, $i$ arbitrary fixed;

is called a *Classical Information Retrieval* (*CIR*).

Two special cases of *CIR* will be defined in the following: Similarity *IR* (equivalent to the classical vector *IR*, Theorem 4.2), and Probabilistic *IR* (equivalent to the traditional probabilistic *IR*, Theorem 4.25).

  Note
The term *similarity IR* instead of *vector IR* is used and suggested because the first one better
  reflects the mathematical properties of this model type. More on this later on. See also part
  1.2.1.

### 2.1.1    Similarity Information Retrieval

*Similarity Information Retrieval* (*SIR*) is defined as a particular case of *CIR*, as follows:

*DEFINITION 4.4*[DOMINICH 1999B]
*Similarity Information Retrieval* (*SIR*) is a *CIR* $\Phi = \langle D, \Re \rangle$ obeying the following two properties S1 and S2:
  S1. $C = 1$;
  S2. $\mu_{a_i}$ is commutative, i.e. $\mu_{a_i}(q, \delta) = \mu_{a_i}(\delta, q)$, $\forall \delta, q \in D$.

The reason for considering property S1 is that in the classical vector model of *IR* there only is one criterion (implicitly or explicitly): relevance. Property S2 reflects that the order in which two objects are considered (when calculating the association measure) is indifferent. We will show that *SIR* and the classical vector *IR* are equivalent (Theorem 4.2).

    As an immediate consequence, it is easy to see that Definition 4.4 can be re–formulated as follows (taking into account points 1—7 of part 2.1):

*DEFINITION 4.5*
*Similarity Information Retrieval (SIR)* is a 2–tuple $\Phi = \langle D, \Re \rangle$, $\Re(q) = a_{\alpha_1}$
$= \{\delta | \mu_{a_1}(q, \delta) > \alpha_1\}$ with the following three properties a) through c):
   a) $0 \le \mu_{a_1}(q, \delta) \le 1$, $\forall \delta$, $q \in D$ (normalisation);
   b) $\mu_{a_1}(q, \delta) = \mu_{a_1}(\delta, q)$, $\forall \delta$, $q \in D$ (commutativity);
   c) $q = \delta \Rightarrow \mu_{a_1}(q, \delta) = 1$, $\delta$, $q \in D$ (reflexivity).

Two special cases of *SIR* are the following (corresponding to the classical binary and non–binary vector *IR*, respectively):

*DEFINITION 4.6*
*Binary Similarity Information Retrieval (BSIR)* is an *SIR* $\Phi = \langle D, \Re \rangle$ with $S = \{0,1\}$.

*DEFINITION 4.7*
*Non–Binary Similarity Information Retrieval (NBSIR)* is an *SIR* $\Phi = \langle D, \Re \rangle$ with $S = [0,1]$.

### 2.1.2   Probabilistic Information Retrieval

Probabilistic Information Retrieval (*PIR*) is defined as another particular case of *CIR*, as follows:

*DEFINITION 4.8* [DOMINICH 1999B]
*Probabilistic Information Retrieval (PIR)* is an *CIR* $\Phi = \langle D, \Re \rangle$ satisfying the following condition P:
   P. $C = 2$

The reason for taking $C = 2$ is that in the traditional probabilistic model of *IR* there are two criteria: relevance and non–relevance. We will show that *PIR* as defined in Definition 4.8 and the classical probabilistic *IR* model are equivalent (Theorem 4.25).
   As an immediate consequence it is easy to see that the above Definition 4.8 can be re–formulated as follows (taking into account points 1—7:

*DEFINITION 4.9*
*Probabilistic Information Retrieval (PIR)* is an *CIR* $\Phi = \langle D, \Re \rangle$ where $C = 2$ and $\Re(q) = \{\delta | \mu_{a_i}(q, \delta) \ge \mu_{a_j}(q, \delta)$, $j = i + (-1)^{i+1}$, $\mu_{a_i}(q, \delta) > \alpha_i \}$.

Two particular *PIR* models, corresponding to the traditional binary and non–binary probabilistic *IR*, are obtained depending on the set *S*.

DEFINITION 4.10
*Binary Probabilistic Information Retrieval (BPIR)* is an *PIR* with S = {0, 1}.

DEFINITION 4.11
*Non–Binary Probabilistic Information Retrieval (NBPIR)* is an *PIR* with S = [0, 1].

As a noteworthy property we shall also show that *SIR* can be conceived as a special case of *PIR* by assuming the properties of symmetry (i.e., there only is one criterion) and commutativity (which does not necessarily hold for *PIR*).

The following schematic shows the structure of the classical *IR* models as derived from the unified definition *CIR*:

| CIR (Definition 4.3) | | | |
|---|---|---|---|
| SIR (Definition 4.4) (equivalent to traditional vector model, Theorem 4.2) | | PIR (Definition 4.8) (equivalent to traditional probabilistic model, Theorem 4.25) | |
| Binary SIR (Definition 4.6) | Non–binary SIR (Definition 4.7) | Binary PIR (Definition 4.10) | Non–binary PIR (Definition 4.11) |

## 3.    SIMILARITY (VECTOR SPACE) INFORMATION RETRIEVAL

In this part it will be shown that *SIR* as a special case of *CIR* is equivalent to the traditional vector *IR* model. Also the foundations for a consistent mathematical theory of *SIR* will be developed, naturally integrating major mathematical results achieved so far in this respect.

Retrieval is based on calculating for any pair of objects (object query, object document) a finite real number expressing a likeness between them. This number has the following properties:

— it is positive;
— it does not depend on the order in which the objects are considered;
— it is maximal when computed for identical objects.

As seen in part 1.2 (relations 3.21, 3.22), these make the following definition possible:

DEFINITION *4.12* [VAN RIJSBERGEN 1979]
A function $\sigma: D \times D \to [0, 1]$ is called a *similarity* if the following properties hold:
a) $0 \leq \sigma (a,b) \leq 1$, $\forall a, b \in D$, normalisation;
b) $\sigma (a, b) = \sigma (b,a)$, $\forall a, b \in D$, symmetry;
c) $a = b \Rightarrow \sigma (a, b) = 1$, $a, b \in D$, reflexivity.

By linking the set $D$ of objects and similarity $\sigma$, the following structure is obtained:

DEFINITION *4.13* [DOMINICH 2000A]
Let $D$ be a set of objects and $\sigma$ a similarity on $D$. A *similarity space* (or $\sigma$–*space*) on $D$ is a 2–tuple $\Sigma = \langle D, \sigma \rangle$.

The vector *IR* model can also be defined defined by specifying retrieval using similarity (see also relation 3.22).

DEFINITION *4.14* [VAN RIJSBERGEN 1979]
Let $\Sigma = \langle D, \sigma \rangle$ be a $\sigma$–space. *Vector Information Retrieval (VIR)* on $D$ is a 3–tuple $\Psi = \langle D, \mathcal{R}, \sigma \rangle$ where $\mathcal{R}(d) = \{x \in D | \sigma(d, x) > \tau, \tau \in \mathbf{R}\}$.

It can be seen that Definition 4.14 is equivalent to Definition 4.5, and thus to Definition 4.4 of *SIR* (similarity $\sigma$ corresponds to membership function $\mu$, and threshold $\tau$ to the cut value $\alpha_1$). It can now be shown that the *SIR* and the classical vector *IR* are equivalent:

**THEOREM 4.2 [DOMINICH 2000B]**
*VIR* and *SIR* are equivalent.

**Proof.**
'$\Leftarrow$', i.e. given *SIR* results in *VIR*. $C = 1$, one criterion, by S1. Let $\tilde{a}_1$ be this criterion and called 'similar'. Because it is the only criterion we can omit it. By S2 we have: $\mu(q, \delta) = \mu(\delta, q)$, $\forall \delta, q \in D$ (commutativity). By property P1 one can write: $q = \delta \Rightarrow \mu(q, \delta) = 1$, $\forall \delta, q \in D$ (reflexivity). Because $\mu$ is normalized we have: $0 \leq \mu(q, \delta) \leq 1$, $\forall \delta, q \in D$. Thus $\mu$ satisfies the conditions a) through c), and hence it is a similarity. Property P2 becomes: $\mathcal{R}(q) = \{\delta | \mu(q, \delta) > \alpha\}$ which is identical to the retrieval condition of *VIR*.

'$\Rightarrow$', i.e. given *VIR* result in *SIR*. Similarity $\sigma$ is a degree to which $b$ is similar to $a$, thus $\sigma(a, b)$ is conceived as being $\mu\tilde{a}_1(a, b)$ where $\tilde{a}_1$ denotes 'similar'. This is equivalent to  S1, i.e. one criterion. Condition a) is equivalent to normalization of point 5), condition b) to  S2, whereas condition c) to property P1. The retrieval condition becomes: $\mathfrak{R}(q) = \{d \in D | \sigma(d, q) > \tau, \tau \in \mathbf{R}\} = \{d \in D | \mu\tilde{a}_1(d, q) = \max \mu\tilde{a}_1(d, q)\} \cap a_\tau.$ ♦

Thus, because *SIR* is a special case of *CIR*, and it is equivalent to the classical vector *IR* model, this latter is itself a special of *CIR*.

The underlying mathematical structures in *SIR* are topological and metric spaces [see, e.g., Raghavan and Wong [1986], Raghavan, Wong and Ziarko [1990], Everett and Cater [1992], Raghavan and Bollmann–Sdorra, [1993], Caid, Dimais and Galiant [1995], Kang and Choi [1997], Kowalski [1997], Ciaccia, Patella and Zezula  [1998], Egghe and Rousseau [1997, 1998]. Everett and Cater, [1992]; Dominich, [1999b, 2000a,b]. The basic results, achieved so far, will be naturally integrated into the present framework of $\sigma$–spaces.

Note

Because of the fundamental role palyed by the concept of $\sigma$–space, it should parhaps gain a mathematical status.

The following theorems express an immediately and naturally arising connection between *SIR* and metrics and topologies.

Noting that (see, e.g., Everett and Cater, [1992]) for every pseudometric $\mu'$ there exists an equivalent $\mu = \mu'/(1 + \mu')$ bounded by 1 (other choices: $\mu = \min(1, \mu')$, $\mu = \sup_E \mu'$), an immediate and naturally arising connection between a pseudo–metric space and a $\sigma$–space is expressed in the following:

**THEOREM 4.3 [DOMINICH 1999B]**
Let $\langle E, \mu \rangle$ be a pseudo–metric space. Then $\langle E, 1 - \mu \rangle$ is a $\sigma$–space.

**Proof.** Let $\sigma = 1 - \mu$. Because $\mu$ is a pseudo–metric, it satisfies the condition:

$$0 \leq \mu(a, b) < +\infty, \ \forall a, b \in E$$

from which it follows that

$$0 \leq \sigma(a, b) \leq 1, \ \forall a, b \in E.$$

$\mu$ also  satisfies the condition

$$\mu(a, b) = \mu(b, a), \quad \forall a, b \in E$$

from which it follows that

$$\sigma(a, b) = \sigma(b, a), \quad \forall a, b \in E.$$

Finally, $\mu$ also satisfies

$$a = b \Rightarrow \mu(a, b) = 0$$

from which we have that

$$a = b \Rightarrow \sigma(a, b) = 1, \quad a, b \in E.$$

Hence, $\sigma$ is a similarity and consequently $\langle E, \sigma = 1 - \mu \rangle$ is a $\sigma$–space. ♦

Theorem 4.3 makes the following definition possible:

*DEFINITION 4.15*
Let $\langle E, \mu \rangle$ be a pseudo–metric space. Then $\langle E, 1 - \mu \rangle$ is the $\sigma$–space *induced* by pseudo–metric $\mu$.

As recent research has revealed, topologies are the underlying mathematical structure of SIR (see, e.g., Everett and Cater, [1992], Egghe and Rousseau [1997, 1998]). In the following, based on these results natural connections with topologies will be introduced.

An immediate connection between *SIR* and a topological space can be established, as follows:

**THEOREM 4.4 [DOMINICH 1999B]**
Let $\langle E, \mu \rangle$ be a pseudo–metric space. Then the induced topological space is an *SIR* on $E$.

**Proof.** Because $\langle E, \mu \rangle$ is a pseudometric space, the $\varepsilon$–balls

$$V(d, x) = \{x | \mu(d, x) < \varepsilon \}, \quad \forall \varepsilon \in \mathbf{R}_+,$$

for an arbitrary element $d \in E$ constitute a base of open sets for a topology with which $E$ is a topological space. On the other hand, the pseudo–metric $\mu$ induces a $\sigma$–space on $E$ with $\sigma = 1 - \mu$. Further we have:

$$V(d, x) = \{x | \mu(d, x) < \varepsilon \}$$

$$= \{x|\ 1 - \sigma(d, x) < \varepsilon\}$$

$$= \{x|\ 1 - \varepsilon < \sigma(d, x)\}$$

$$= \{x\ |\ \sigma(d, x) > \tau\} = \mathfrak{R}(d)$$

where $\tau = 1 - \varepsilon$. Hence $\langle E, \mu \rangle$ is an *SIR*. ♦

Moreover, a connection between *SIR* and a Hausdorff space can also established at a level of equivalence classes.

For technical reasons the following theorems are needed first:

**THEOREM 4.5 [CSÁSZÁR 1978]**
Let $\langle E, \mu \rangle$ be a pseudo–metric space. Then the relation $\sim$, $x \sim y \Leftrightarrow \mu(x, y) = 0$, $\forall x, y \in E$, is an equivalence relation on $E$.

**Proof.**
$\mu(x, x) = 0 \Rightarrow x \sim x$, i.e., the relation $\sim$ is reflexive. $\mu(x, y) = \mu(y, x) \Rightarrow (x \sim y \Rightarrow y \sim x)$, i.e., the relation $\sim$ is symmetric. $x \sim y \wedge y \sim z \Rightarrow \mu(x, y) = 0 \wedge \mu(y, z) = 0 \Rightarrow 0 \le \mu(x, z) \le \mu(x, y) + \mu(y, z) = 0 + 0 \Rightarrow \mu(x, z) = 0 \Rightarrow x \sim z$, i.e. relation $\sim$ is transitive. ♦

The relation $\sim$ defines a partition on $E$. Let $E^*$ denote the set of equivalence classes. Then:

**THEOREM 4.6 [CSÁSZÁR 1978]**
Let $\langle E, \mu \rangle$ be a pseudo–metric space, and $\sim$ relation $x \sim y \Leftrightarrow \mu(x, y) = 0$, $\forall x, y \in E$. Then the space $\langle E^*, \mu^* = \mu \rangle$ is a metric space ($\mu^*(A, B) = \mu(x, y)$, $A, B \in E^*$, $x \in A, y \in B$).

**Proof.** Firstly, we prove that the value of $\mu^*$ does not depend on the choice of elements.
Consider $a, b \in A \in E^*$ and $c, d \in A \in E^*$. One can write:

$$\mu(a, c) \le \mu(a, b) + \mu(b, c)$$

$$\le 0 + \mu(b, d) + \mu(d, c)$$

$$= 0 + \mu(b, d) + 0 = \mu(b, d)$$

and

$$\mu(b, d) \le \mu(b, a) + \mu(a, d)$$

$$\le 0 + \mu(a, c) + \mu(c, d)$$

$$= 0 + \mu(a, c) + 0$$

$$= \mu(a, c)$$

Hence $\mu(a, c) = \mu(b, d)$.

Further, it can be easily shown that $\mu^*$ is a metric. Hence $\langle E^*, \mu^* \rangle$ is a metric space. ♦

As consequences the following connections between a Hausdorff space and *SIR* can now be easily established:

**COROLLARY 4.6.A [DOMINICH 2000B]**
The Hausdorff space induced by the metric $\mu^*$ is an *SIR*.

**Proof.** $\mu^*$ being a metric, the induced topological space on E* is a Hausdorff space. ♦

**COROLLARY 4.6 B [DOMINICH 2000B]**
Let $S = \langle E, \sigma \rangle$ be a $\sigma$–space. If $\delta = 1 - \sigma$ is a pseudo–metric/metric on $E$, then the induced topological space/Hausdorff space on $E$ and *SIR* on $E$ are equivalent.

**Proof.** Immediate. ♦

The idea of separation expressed in a Hausdorff space (Corollary 4.6A and 4.6B) can also be found in [Everett and Cater, 1992] as follows.

A special similarity is defined as follows (analogous to an inverse of the simple matching coefficient (3.16) in the sense that it is a measure of how much $x$ and $y$ have in common):

$$\rho(d, x) = |\sigma(d, x) - \sigma(d, y)| \tag{4.3}$$

If this is finite then let

$$\text{dist}(x, y) = \sup_d \{|\sigma(d, x) - \sigma(d, y)|\} \tag{4.4}$$

Let two objects $x$ and $y$ be equivalent (relative to a fixed $d$) if $\sigma(d, x) = \sigma(d, y)$. Thus, an equivalence relation can be defined over $D$. Let $D^*$ the quotient space under this equivalence relation, i.e., the set of equivalence classes. Then it can be shown that:

**THEOREM 4.7 [EVERETT AND CATER 1992]**
The mapping

$$\text{dist}^*: D^* \times D^* \to \mathbf{R}_+, \quad \text{dist}^*(a^*, b^*) = \text{dist}(a, b)$$

is a metric on $D^*$, and hence induces a metric topology on $D^*$.

**Proof.** Similar to the proof of Theorem 4.6 ♦

Let

$$\text{dist}'(x, y) = \sup_x \frac{\rho(d, x)}{1 + \rho(d, x)} \tag{4.5}$$

and

$$\text{dist}''(x, y) = \sup_x (\min(1, \rho(d, x))) \tag{4.6}$$

be other two measures (normalised versions of $\rho$; see also previous notes on $\mu$ and $\mu'$). As an interesting property, it can be shown that:

**THEOREM 4.8 [EGGHE AND ROUSSEAU 1998]**
dist, dist' and dist" are equivalent.

**Proof.** Let $B(d, \varepsilon)$, $B'(d, \varepsilon)$ and $B''(d, \varepsilon)$, $\forall \, \varepsilon > 0$, $\forall \, d \in D$, be the dist, dist' and dist", respectively, $\varepsilon$–balls around $d$.
   Because $\text{dist}'(x, y) \le \text{dist}''(x, y)$ we have $B''(d, \varepsilon) \subseteq B'(d, \varepsilon)$.
Let $\varepsilon' = \varepsilon/(2 + \varepsilon)$. Then:

$$\text{dist}'(x, y) < \varepsilon' \implies \frac{\rho(d, x)}{1 + \rho(d, x)} < \varepsilon'$$

Hence $\rho(d, x) < \varepsilon/2$. So $dist''(x, y) < \varepsilon$, and thus $B'(d, \varepsilon') \subseteq B''(d, \varepsilon)$. This means that $dist'$ and $dist''$ generate the same topology. Since $dist'(x, y) \leq dist(x, y)$ we have $B(d, \varepsilon') \subseteq B'(d, \varepsilon)$. Further $dist'(x, y) < \varepsilon'$ and so $\rho(d, x) < \varepsilon/2$. Hence $dist(x, y) < \varepsilon$ and thus $B'(d, \varepsilon') \subseteq B(d, \varepsilon)$. ♦

## 3.1    Binary Similarity Information Retrieval

This part develops the foundations for a mathematical theory of the binary *SIR*, as an application.

Let

$$T = \{t_1, ..., t_i, ..., t_N\}, \quad N \geq 1 \tag{4.7}$$

be a set of *identifiers*. Each object is conceived as

$$d_j \in D = \{d_j|\, d_j \in \wp(T), j = 1, ..., M\}, \quad M > 1 \tag{4.8}$$

is associated a binary vector as follows:

$$f: D \rightarrow \{0, 1\}^N \tag{4.9}$$

$$f(d_j) = \mathbf{x}^j = (x_{j1}, ..., x_{jk}, ..., x_{jN}) \tag{4.10}$$

$$x_{jk} = 1, \text{ if } t_k \in d_j \text{ and } x_{jk} = 0, \text{ if } t_k \notin d_j; k = 1, ..., N \tag{4.11}$$

Thus an *IRF* $\Phi = \langle D, \mathfrak{R} \rangle$ is associated with a *binary frame*

$$\langle X = \{\mathbf{x}^j|\, f(d_j) = \mathbf{x}^j, j = 1, ..., M\}, \mathfrak{R} \rangle \tag{4.12}$$

It may happen that two different objects have the same binary description. This property makes the following definition possible:

*DEFINITION 4.16* [WONG AND YAO 1990]
Objects $d_i$ and $d_j$ are *equivalent*, denoted $\approx$, if they have the same binary description, i.e. $d_i \approx d_j \Leftrightarrow f(d_i) = f(d_j)$.

It is easy to see that:

**THEOREM 4.9**

The relation $\approx$ is an equivalence relation on $D$.

**Proof.**

a) $\approx$ is reflexive, i.e.,

$$d_i \approx d_i : f(d_i) = f(d_i)$$

b) $\approx$ is symmetric, i.e.,

$$d_i \approx d_j \Rightarrow d_j \approx d_i : d_i \approx d_j \Rightarrow f(d_i) = f(d_j) \Rightarrow d_j \approx d_i$$

c) $\approx$ is transitive, i.e.,

$$d_i \approx d_j \wedge d_j \approx d_k$$

$$\Rightarrow d_i \approx d_k : d_i \approx d_j \wedge d_j \approx d_k$$

$$\Leftrightarrow f(d_i) = f(d_j) \wedge f(d_j) = f(d_k)$$

$$\Rightarrow f(d_i) = f(d_k) \Rightarrow d_i \approx d_k \quad \blacklozenge$$

The relation $\approx$ defines a partition of $D$:

$$D = \bigcup_r C_r, \; C_r = \{d \in D \mid d \sim r, r \in D\} \tag{4.13}$$

where $r$ is a class representative, $C_r \cap C_p = \varnothing, \forall r \neq p$. This yields the following:

**THEOREM 4.10**

If function $f$ is surjective, the number of equivalence classes is equal to $2^N$.

**Proof.** $f$ being surjective we have

$$\forall \mathbf{x} \in \{0, 1\}^N \; \exists d \in D \text{ such that } f(d) = \mathbf{x}$$

i.e., all objects have distinct binary descriptions. Hence the number of equivalence classes is equal to the number of elements, i.e. cardinality, of set $\{0, 1\}^N$ which is $|\{0, 1\}^N| = 2^N$ $\blacklozenge$

Theorem 4.10 makes the following definition of a discrimination power or capacity possible (i.e. the total number of distinct representations):

*DEFINITION 4.17*
The *discrimination power* (*d–power*) of a binary frame $\langle X, \mathfrak{R} \rangle$ is $2^N$.

Further, based on Definition 4.17, the following equivalence relation can be defined:

*DEFINITION 4.18*
Two binary frames, $X_1$ and $X_2$, are *capacity equivalent* (*or c–equivalent*) if their *d*–powers are equal to each other.

Notice that $X_1$ and $X_2$ have equal *d*–powers if the corresponding sets $T_1$ and $T_2$ are of equal cardinalities, i.e., $|T_1| = |T_2|$. It is easy to check that the *d*–power is an equivalence relation. The *d*–power relation makes it possible to define an ordering relation:

*DEFINITION 4.19*
A binary frame $X_1$ with *d*–power $K_1$ is *more refined* than a binary frame $X_2$ with *d*–power $K_2$ if $K_1 \geq K_2$. Notation: $X_1 \prec X_2 \Leftrightarrow K_1 \geq K_2$.

As a consequence, because the above relation $\prec$ is reflexive, transitive and antisymmetric, the set of binary frames is an ordered set.

Function $f$ (3.10) above can be used to express the cardinality of an object $d_j$:

$$|d_j| = x_{j1} + \ldots + x_{jk} + \ldots + x_{jN} = \sum_{k=1}^{N} x_{jk} \tag{4.14}$$

This allows for expressing the cardinalities of set operations:

**THEOREM 4.11**
Cartesian Product:

$$|d_i \times d_j| = \sum_{k=1}^{N} x_{ik} \sum_{k=1}^{N} x_{jk} \tag{4.15}$$

Intersection:

$$|d_i \cap d_j| = \sum_{k=1}^{N} x_{ik} x_{jk} = (\mathbf{x}^i, \mathbf{x}^j) \tag{4.16}$$

Difference:

$$|d_i \setminus d_j| = \sum_{k=1}^{N} x_{ik} (1 - x_{jk}) = (\mathbf{x}^i, 1 - \mathbf{x}^j) \tag{4.17}$$

Union:

$$|d_i \cup d_j| = \sum_{k=1}^{N} (x_{ik} + x_{jk}) / 2^{x_{ik} x_{jk}} \tag{4.18}$$

Symmetric difference:

$$|d_i \Delta d_j| = \sum_{k=1}^{N} x_{ik} + \sum_{k=1}^{N} x_{jk} - 2 \sum_{k=1}^{N} x_{ik} x_{jk} \tag{4.19}$$

**Proof.**

Cartesian Product: $|d_i \times d_j| = |d_i| \cdot |d_j| = \sum_{k=1}^{N} x_{ik} \sum_{k=1}^{N} x_{jk}$ .

Intersection: $|d_i \cap d_j| = |\{t_k | t_k \in d_i \wedge t_k \in d_j\}| = \sum_{k=1}^{N} x_{ik} x_{jk}$ .

Difference: $|d_i \setminus d_j| = |\{t_k | t_k \in d_i \wedge t_k \notin d_j\}| = \sum_{k=1}^{N} x_{ik}(1 - x_{jk})$ .

Union: $|d_i \cup d_j| = |\{t_k | t_k \in d_i \vee t_k \in d_j\}| = \sum_{k=1}^{N} (x_{ik} + x_{jk}) / 2^{x_{ik} x_{jk}}$

Symmetric difference:

$$|d_i \, \Delta \, d_j| = |(d_i \setminus d_j) \cup (d_j \setminus d_i)| = |d_i \setminus d_j| + |d_j \setminus d_i| = \sum_{k=1}^{N} (x_{ik} + x_{jk} - 2x_{ik}x_{jk}) \quad \blacklozenge$$

The following is a typical association measure (known as Dice's coefficient, see also 3.18) widely used:

$$m_1(d_i, d_j) = 2 \cdot \frac{|d_i \cap d_j|}{|d_i| + |d_j|} \tag{4.20}$$

This rewrites as follows (using binary vectors):

**THEOREM 4.12**

For $\mathbf{x}^i$ and $\mathbf{x}^j$ not both identically null:

$$m_1(d_i, d_j) = 2 \cdot \frac{\displaystyle\sum_{k=1}^{N} x_{ik}x_{jk}}{\displaystyle\sum_{k=1}^{N} (x_{ik} + x_{jk})} \tag{4.21}$$

**Proof**. Immediate. $\blacklozenge$

**THEOREM 4.13**

$m_1$ is a similarity.

**Proof.** It is obvious that $m_1(d_i, d_j) = m_1(d_j, d_i)$ and $m_1(d_i, d_j) \geq 0$. Further one can write:

$$x_{ik} + x_{jk} \geq 2x_{ik}x_{jk}, \quad \forall k = 1, ..., N \implies m_1(d_i, d_j) \leq 1$$

Also

$$\mathbf{x}^i = \mathbf{x}^j \implies 2 \sum_{k=1}^{N} x_{ik}x_{jk} = \sum_{k=1}^{N} 2x_{ik}$$

$$\sum_{k=1}^{N} (x_{ik} + x_{jk}) = \sum_{k=1}^{N} 2x_{ik} \implies m_1(\mathbf{x}^i, \mathbf{x}^i) = 1 \quad \blacklozenge$$

The set $X$ of all vectors $\mathbf{x}^j$, $j = 1, ..., M$, can be viewed as a metric space:

**THEOREM 4.14**
The space $\langle X, \delta \rangle$ with

$$\delta(\mathbf{x}^u, \mathbf{x}^v) = \frac{\displaystyle\sum_{k=1}^{N} x_{uk} + \sum_{k=1}^{N} x_{vk} - 2 \sum_{k=1}^{N} x_{uk} x_{vk}}{\displaystyle\sum_{k=1}^{N} x_{uk} + \sum_{k=1}^{N} x_{vk}}$$

is a metric space.

**Proof.**
a) We prove that $0 \leq \delta(\mathbf{x}^u, \mathbf{x}^v) < +\infty$. One can write:

$$x_{uk} + x_{vk} \geq 2 x_{uk} x_{vk}, \ \forall k = 1, ..., N \Rightarrow \sum_{k=1}^{N} (x_{uk} + x_{vk}) \geq \sum_{k=1}^{N} x_{uk} x_{vk}$$

hence $\delta$ is positive.
b) We prove that if $\mathbf{x}^u = \mathbf{x}^v$ then $\delta(\mathbf{x}^u, \mathbf{x}^v) = 0$. We have:

$$\mathbf{x}^u = \mathbf{x}^v \Rightarrow x_{uk} = x_{vk}, \ \forall k, \text{ furthermore } x_{uk} x_{vk} = (x_{uk})^2 = x_{uk} \text{ as binary values}$$

hence

$$\sum_{k=1}^{N} x_{uk} + \sum_{k=1}^{N} x_{vk} - 2 \sum_{k=1}^{N} x_{uk} x_{vk} = 2 \sum_{k=1}^{N} x_{uk} - 2 \sum_{k=1}^{N} (x_{uk})^2 = 0$$

and thus $\delta(\mathbf{x}^u, \mathbf{x}^v) = 0$.
c) It is immediate that $\delta(\mathbf{x}^u, \mathbf{x}^v) = \delta(\mathbf{x}^v, \mathbf{x}^u)$.
d) We prove that if $\delta(\mathbf{x}^u, \mathbf{x}^v) = 0$ then $\mathbf{x}^u = \mathbf{x}^v$. we have:

$$\delta(\mathbf{x}^u, \mathbf{x}^v) = 0$$

$$\Rightarrow \sum_{k=1}^{N} x_{uk} + \sum_{k=1}^{N} x_{vk} = 2 \sum_{k=1}^{N} x_{uk} x_{vk}$$

$$\Rightarrow \sum_{k=1}^{N} (x_{uk} + x_{vk} - 2x_{uk}x_{vk}) = 0$$

$$\Rightarrow x_{uk} + x_{vk} - 2x_{uk}x_{vk} = 0, \ \forall k$$

terms under summation being non–negative $\Rightarrow x_{uk} = x_{vk}, \ \forall k \Rightarrow \mathbf{x}^u = \mathbf{x}^v$.

e) Prove the triangle inequality: $\delta(\mathbf{x}^i, \mathbf{x}^j) \le \delta(\mathbf{x}^i, \mathbf{x}^u) + \delta(\mathbf{x}^u, \mathbf{x}^j)$. This rewrites as:

$$1 - 2 \frac{\displaystyle\sum_{k=1}^{N} x_{ik}x_{jk}}{\displaystyle\sum_{k=1}^{N} x_{ik} + \sum_{k=1}^{N} x_{jk}}$$

$$\le \ 1 - 2 \frac{\displaystyle\sum_{k=1}^{N} x_{ik}x_{uk}}{\displaystyle\sum_{k=1}^{N} x_{ik} + \sum_{k=1}^{N} x_{uk}} + 1 - 2 \frac{\displaystyle\sum_{k=1}^{N} x_{uk}x_{jk}}{\displaystyle\sum_{k=1}^{N} x_{uk} + \sum_{k=1}^{N} x_{jk}}$$

$$\Rightarrow 0.5 \ge \frac{\displaystyle\sum_{k=1}^{N} x_{ik}x_{jk}}{\displaystyle\sum_{k=1}^{N} x_{ik} + \sum_{k=1}^{N} x_{jk}} + \frac{\displaystyle\sum_{k=1}^{N} x_{ik}x_{uk}}{\displaystyle\sum_{k=1}^{N} x_{ik} + \sum_{k=1}^{N} x_{uk}} - \frac{\displaystyle\sum_{k=1}^{N} x_{uk}x_{jk}}{\displaystyle\sum_{k=1}^{N} x_{uk} + \sum_{k=1}^{N} x_{jk}}$$

This holds because the maximum of the last right–hand side is ½. ♦

It can now be proved that the above metric induces an *SIR* in which retrieval is based on the typical association measure $m_1$.

**THEOREM 4.15**
Let $D$ be a set of objects with their associated binary vector representations, $X$, and metric $\delta$ above. Then the Hausdorff space $\langle X, \delta \rangle$ and the *SIR* $\langle X, m_1 \rangle$ are equivalent.

**Proof.**
Prove that given $\delta$ results in *SIR*. Because $\langle X, \delta \rangle$ is a metric space, it follows that $\langle X, \Re, m_1 \rangle$ is an *SIR* as it is a Hausdorff space.

Prove that given an *SIR* $\langle X, \Re, m_1 \rangle$ results in defining a Hausdorff space on $X$. Notice that $\delta = 1 - m_1$ is a metric, hence the space $\langle X, \delta \rangle$ is a metric space. ◆

Other properties are:

- $0 \leq \delta^2 + \sigma^2 \leq 1$
- $0 \leq \delta\sigma \leq \frac{1}{2}$
- $(\delta + \sigma)^2 = 1$

Other widely used association measure is the cosine measure (see also 3.17):

$$m_3(d_i, d_j) = \frac{|d_i \cap d_j|}{|d_i|^{1/2} \cdot |d_j|^{1/2}} \tag{4.22}$$

This re–writes as:

**THEOREM 4.16**

$$m_3(d_i, d_j) = \frac{\sum_{k=1}^{N} x_{ik}x_{jk}}{(\sum_{k=1}^{N} x_{ik})^{1/2}(\sum_{k=1}^{N} x_{jk})^{1/2}} \tag{4.23}$$

**Proof.** Immediate. ◆

It can be easily seen that:

**THEOREM 4.17**
$m_3$ is a similarity.

**Proof.** It is immediate that $m_3(d_i, d_j) = m_3(d_j, d_i)$ and $m_3(d_i, d_j) \geq 0$. Further:

$m_3(d_i, d_j) \leq 1$ by the Hölder inequality ◆

As an interesting property, the following holds:

**THEOREM 4.18**
The following inequality holds: $m_1 \leq m_3$

**Proof.**
We have the following sequence of implications:

$$\left(\sum_{k=1}^{N} x_{ik} - \sum_{k=1}^{N} x_{jk}\right)^2 \geq 0$$

$$\Rightarrow \left(\sum_{k=1}^{N} x_{ik}\right)^2 + \left(\sum_{k=1}^{N} x_{jk}\right)^2 + 2\sum_{k=1}^{N} x_{ik} \sum_{k=1}^{N} x_{jk} - 4\sum_{k=1}^{N} x_{ik} \sum_{k=1}^{N} x_{jk} \geq 0$$

$$\Rightarrow \left(\sum_{k=1}^{N} x_{ik} + \sum_{k=1}^{N} x_{jk}\right)^2 \geq 4\sum_{k=1}^{N} x_{ik} \sum_{k=1}^{N} x_{jk}$$

$$\Rightarrow \sum_{k=1}^{N} (x_{ik} + x_{jk}) / 2 \geq \left(\sum_{k=1}^{N} x_{ik}\right)^{1/2} \left(\sum_{k=1}^{N} x_{jk}\right)^{1/2}$$

$$\Rightarrow m_1 \leq m_3 \blacklozenge$$

A special (and undesirable) case is when no objects are retrieved. The following definition is introduced:

*DEFINITION 4.20*
Let $\langle X, \mathfrak{R}, \sigma \rangle$ be a binary frame. A *no–hit* is a binary vector **v** such that $\sigma(\mathbf{x}^j, \mathbf{v}) = 0$, $\forall j = 1, ..., M$.

The total number of possible no–hits is expressed in the following:

**THEOREM 4.19**
The number of possible no–hits is equal to $2^Z$, where $Z$ denotes the number of 0–columns in matrix $\chi = (\mathbf{x}^j)_{j=1,...,M}$

**Proof.** A no hit $\mathbf{v}$ satisfies the simultaneous system of equations

$$\sigma(\mathbf{x}^j, \mathbf{v}) = 0, \ j = 1, ..., M$$

This is equivalent to

$$(\mathbf{v}, \mathbf{x}^j) = 0, j = 1, ..., M$$

which rewrites as the (linear) matrix equation

$$\chi \times \mathbf{v}^T = 0$$

hence

$$x_{jk} v_k = 0, \ j = 1, ..., M, \ k = 1, ..., N$$

From this it follows that if $x_{jk} = 0$ then $v_k \in \{0,1\}$. Thus, the number of such vectors $\mathbf{v}$ is equal to the number of binary arrangements on $Z$ positions which is equal to $2^Z$. ♦

See Appendix 1 for a complex example and complete procedures (in MathCAD 8 Plus Professional).

### 3.1.1    Conclusions

Whatever the number $M$ of documents, it is impossible to represent more than $2^N$ documents. In other words, the capacity of the Binary *SIR* is determined solely by the number $N$ of index terms, albeit to generate all possible representations is algorithmically intractable (**NP** Class problem, because of the exponential time complexity).

Typically, conceiving all possible retrievals, the Binary *SIR* may be viewed as a metric/topological/Hausdorff space.

The Binary *SIR* is, at least in principle or formally, predictable or deterministic in that all possible no–hits, queries and retrived documents can be calculated in advance (once $D$, $T$ and $\sigma$ are given).

Properties P1 and P2 of Definition 4.3 may be conceived as axioms, too. They are independent and also consistent (an interpretation is as follows: $T = $ keywords, $D = O$, $D_j = \{o_j = \delta_j \in \wp(O)\}$, $\mu_{\delta_j}(t_k) = 1$ if $t_k \in o_j$ and 0 otherwise, $\mu_{a_i} = $ Dice's coefficient). Also, the Binary *SIR* above is another noncontradictory interpretation. P1 and P2 are not complete as there may be different non–isomorphic interpretations (e.g. binary, non–binary).

## 3.2    Non–Binary Similarity Information Retrieval

This part develops the foundations for a mathematical theory of non–binary *SIR*, as an application.

Each object $d_j \in D$ is identified with an $N$–dimensional vector as follows:

$$f: D \to [0, 1]^N \tag{4.24}$$

$$f(d_j) = \mathbf{w}^j = (w_{j1}, ..., w_{ji}, ..., w_{jN}) \tag{4.25}$$

Let us recall a few examples for $w_{ji}$ (see also part 1.2):

$$\text{number\_of\_occurrences\_of\_}i\text{th\_term} \tag{4.26}$$

$$\text{frequency} = \frac{\text{number\_of\_occurrences\_of\_}i\text{th\_term}}{\text{total\_number\_of\_terms\_in\_document\_}d_j} \tag{4.27}$$

$$\text{inverse document frequency } w_{ij}: \quad w_{ij} = -\log_2(df_i/M) \tag{4.28}$$

where $M$ is the total number of documents, i.e. $|D| = M$, and $df_i$ is the number of documents in which term $t_i$ occurs. Thus, an *IRF* $\langle D, \Re \rangle$ is associated a non–binary frame $\langle W = \{\mathbf{w}^j | j = 1, ..., M\}$.

A $d$–power (similar to that for *BSIR*) cannot be defined, more presicely it would be infinite.

A typical similarity function is the cosine measure (which is a generalisation of $m_3$, 4.23):

$$s(\mathbf{w}^u, \mathbf{w}^v) = \frac{(\mathbf{w}^u, \mathbf{w}^v)}{\|\mathbf{w}^u\| \, \|\mathbf{w}^v\|} \tag{4.29}$$

It can be easily seen that $s(\mathbf{w}^u, \mathbf{w}^v)$ satisfies the similarity conditions: it is normalised, symmetric and reflexive.

The cosine measure, as we have seen it in Chapter 3, can be viewed as a normalised version of the dot product. We are going to show that the dot product (recall formula 3.16) itself can be normalised.

Typically, the weights $w_{ij}$ have a statistical nature, i.e. they are ultimately based on term occurrences (in $D_i$ or $Q$). Because the document vectors $\mathbf{v}_i$ are already known (as the documents are given), the question is whether the set of all possible $\mathbf{v}_Q$s is finite. The answer is yes, since all possible $\mathbf{v}_Q$s can, at least in principle, be predicted: they are all the conceivable arrangements of all terms $t_j$, and their number is a finite multiple of $2^N$ (albeit intractable algorithmically). Thus the vector $IR$ has a deterministic nature (see previous part, too). Let $S = \sup_{i,Q}(\mathbf{v}_i, \mathbf{v}_Q)$; $S$ will ensure that normalisation hold. Further let $P = \max_{i=1 \ N} \sum_{j=1}^{M} (w_{ij})^2$; $M$ will ensure that reflexivity holds, i.e., $s_{iQ} = 1$ if $\mathbf{v}_i = \mathbf{v}_Q$. Thus, 3.16 can be re–defined as follows:

$$s'_{iQ} = \frac{s_{iQ}}{P}, \quad \text{if } \mathbf{v}_i = \mathbf{v}_Q$$

$$s'_{iQ} = \frac{s_{iQ}}{S}, \quad \text{if } \mathbf{v}_i \neq \mathbf{v}_Q$$

At this point let us recall two concepts from topology.

The standard topology on $[0, 1]$ is the topology whose basis consists of all the intervals $[0, b)$, $(a, b)$ and $(a, 1]$, where $0 < a, b < 1$. With this topology, the space $\mathbf{I}^n = [0, 1]^n$ is a topological space.

The space $\mathbf{S}^{n-1} = \{\mathbf{x} \in \mathbf{E}^n |\ \|\mathbf{x}\| = 1\}$ is the $(n - 1)$-dimensional sphere, and a topological space, where $\mathbf{E}^n$ is the Euclidean space and $\|\mathbf{x}\|$ the usual norm. Let $(\mathbf{S}^{n-1}, \varepsilon)$ denote the usual Euclidean topology.

As a noteworthy property, it can be shown that:

**THEOREM 4.20 [EVERETT AND CATER 1992]**
The function $cs$ defined as

$$cs = s(\mathbf{w}^u, \mathbf{w}^v), \quad \mathbf{w}^u \neq \mathbf{0}, \ \mathbf{w}^v \neq \mathbf{0}$$

$$cs = 0, \quad \mathbf{w}^u = \mathbf{0} \text{ or } \mathbf{w}^v = \mathbf{0}$$

is discontinuos at $\mathbf{w}^u = \mathbf{0}$ and $\mathbf{w}^v = \mathbf{0}$.

**Proof.** Counterexample: for $n = 1$ we have $s(\mathbf{w}^u, \mathbf{w}^v) = 1$. Or, alternatively: Given any non–zero $\mathbf{w}^u$ and $\mathbf{w}^v$, and let $\mathbf{w}^{u,n} = 1/n\ \mathbf{w}^u$. Then $\mathbf{w}^{u,n} \to \mathbf{0}$, i.e., the limit is the null vector. We have $s(\mathbf{w}^{u,n}, \mathbf{w}^v) = s(\mathbf{w}^u, \mathbf{w}^v)$, $\forall n$. Alternatively, let $M = \min_w s(\mathbf{w}, \mathbf{w}^v)$. Then we can choose $\mathbf{w}^u$ such that $s(\mathbf{w}^{u,n}, \mathbf{w}^v)$ can take any preselected value in the interval $[M, 1]$. Therefore $s(\mathbf{w}^u, \mathbf{w}^v)$ cannot have a limit when $\mathbf{w}^u \to \mathbf{0}$. Similarly for $\mathbf{w}^v$. ◆

Although $1 - s(\mathbf{w}^u, \mathbf{w}^v)$ is not a pseudo–metric (it does not satisfy the triangle inequality, for example for $\mathbf{w}^u = [0, 0.506, 0.068, 0.982, 0.616, 0.368]^T$, $\mathbf{w}^v = [0, 0.642, 0.879, 0.412, 0.964, 0.342]^T$, $\mathbf{w}^z = [0, 0.488, 0.762, 0.336, 0.36, 0.081]^T$), a transition to topology can be established in this case, too, albeit not so smoothly as in the binary case (Theorem 4.14), as follows.

It is obvious that $s(\mathbf{w}^u, \mathbf{w}^v)$ cannot distinguish between objects whose vectors lie on the same straight positive half–line (through the origin). Thus, all such objects (vectors) can be regarded as being equivalent (0 not included). On the other hand, we have just seen (Theorem 4.20) that the cosine measure cannot be extended so as to to be continuous. Because of these two reasons, a connection to topology is established at the level of equivalence classes (taking into account Theorem 4.7, too).

**THEOREM 4.21 [EGGHE AND ROUSSEAU 1998]**
Let $D^*$ denote the set of all equivalence classes, i.e., $D^* = \mathbf{I}^n \setminus \{\mathbf{0}\}$, and let $\sigma$ of (4.4) be $s$ of (4.29). Then $(D^*, \sigma)$ is a Hausdorff space.

**Proof.**
Consider the mapping $f: D^* \to \mathbf{S}$, where $\mathbf{S}$ is the positive part of $\mathbf{S}^{n-1}$. It can be easily seen that $f$ is a homeomorphism, and thus the sapces $(D^*, \sigma)$ and $(\mathbf{S}, \varepsilon)$ are equivalent. Hence, $(D^*, \sigma)$ is a Hausdorff space, too. ◆

## 3.3    Conclusions

As opposed to the Binary *SIR*, in the Non–binary *SIR*, it is impossible to specify a capacity (i.e., how many documents can be represented) as a finite number.

There may be documents having the same non–binary representations depending on the choice of weights.

Typically, conceiving all possible retrievals, the non–Binary *SIR* may be viewed as a Hausdorff space but at a level of equivalent documents.

As opposed to the Binary *SIR*, the non–Binary *SIR* is not deterministic (in the sense that the binary is, as seen).

Another noteworthy property characterising *SIR* in general is that its retrieval function (similarity) is commutative — as opposed to the probabilistic model (in which the retrieval function is not commutative; see next part).

### 3.4      Bibliographical Remarks

Different mathematical (mainly topological) properties of *VIR* are investigated by e.g. Raghavan and Wong [1986], Raghavan, Wong and Ziarko [1990], Everett and Cater [1992], Caid, Dimais and Galiant [1995], Kang and Choi [1997], Kowalski [1997], Ciaccia, Patella and Zezula [1998], Egghe and Rousseau [1997, 1998].

The name of Vector *IR* is a historically developed name and could be misleading. The *VIR* model is called so (Vector *IR*) because objects are represented as strings of real numbers, just as if they were real valued vectors (in a mathematical sense) although their collection does not necessarily form a vector (linear) space (in the mathematical sense). One reason for this is, for example, that there is no guarantee that the sum of two vectors represent another existing document; in other words, addition is not an internal operation, which is a requirement in linear spaces. This is one reason why the concept of Similarity *IR* (*SIR*) is suggested in this book instead of Vector *IR*. Also, *SIR* better reflects the structural properties of this model type [see, e.g., Bollmann–Sdorra and Raghavan, 1993 for an analysis].

Also they argue, using thought or contrived (*gedanken*) examples, that having documents and queries in the same conceptual space may yield meaningless preference, similarity, term independence and linearity. One should note, however, that the two structures studied by Bollmann–Sdorra and Raghavan are not different in principle, i.e., a purely mathematical sense (they both are relational systems with a preference relation), rather, they bear specific differences, e.g., the way in which their preference relations are defined. If one takes two (principially and/or specifically) different conceptual structures, one for documents and another one for queries, the relation or connection or mapping — which may yield a new 'compound' structure — between these two structures should also be analyzed (mathematically).

In this book, the same conceptual space is adopted both for queries and documenst for two reasons: i) *modelling reason* — we are only concerned with purely formal aspects in a mathematical foundation for *IR*, ii) *practical reason* — in practical implementations of *VIR*, documents and queries are treated as objects of the same type.

## 4.      PROBABILISTIC INFORMATION RETRIEVAL

In this part it will be shown that *PIR*, as a special case of *CIR*, is equivalent to the traditional probabilistic *IR* in the case of optimal retrieval. Also, together with Chapter 5, the foundations for a mathematical theory of *PIR* is developed.

The usual definition of the classical probabilistic *IR* is as follows (see part 1.3 and relation 3.23):

*DEFINITION 4.21* [VAN RIJSBERGEN 1979]
Let $D$ be a set of documents, $q \in D$ a query, and $P(R|(q, d))$ and $P(I|(q, d))$ the probability that document $d \in D$ is relevant/irrelevant to query $q$, respectively. Let $\Re(q)$ denote the set of retrieved documents in response to query $q$. A document $d$ is selected in response to a query $q$ if

$$P(R|(q, d)) \geq P(I|(q, d)) \text{ (Bayes' Decision Rule)}$$

i.e.,

$$\Re(q) = \{d | P(R|(q, d)) \geq P(I|(q, d))\}$$

More exactly, $P(R|(q, d))$ and $P(I|(q, d))$ are probabilities *associated* with $d$ when it is judged to be relevant to $q$ and irrelevant, respectively.

The selected documents can be ranked in decreasing order of their relevance ('probability ranking principle'). A cut off value is usually used.

The estimation of $P(R|(q, d))$ and $P(I|(q, d))$ is based on Bayes' Formula, and there is a vast literature and experience on the technicalities of this.

From a purely mathematical point of view one should note the following. For the conditional probability $P(R|(q, d))$ to make sense, $R$ and $(q, d)$ should be homogenous entities, i.e., they should be events of the same $\sigma$–algebra over a field of events $\Omega$ (Kolmogoroff's concept of probability, as these probabilities are of a statistical nature). The same holds for $P(I|(q, d))$). Although the symbol $P(R|(q, d))$ (or $P(I|(q, d))$) is — traditionally — called the probability of relevance $R$ (non–relevance $I$) of document $d$ with respect to query $q$, it is in fact a probability that is *assigned* to document $d$ so as to express a degree of relevance (irrelevance) to query $q$.

Let $D$ be a set of objects, $q \in D$ an arbitrary fixed object and $\tilde{a}_1$ and $\tilde{a}_2$ two *criteria* called *relevant* and *non–relevant*, respectively. Let $\mu_{\tilde{a}_i}(q, d)$, $i = 1, 2$, denote a degree to which an arbitrary object $d \in D$ satisfies criterion $\tilde{a}_i$ relative to $q$.

Probabilistic Information Retrieval (*PIR*) is now formally defined as a(nother) special case of *CIR*, as follows (see also Definition 4.9 taking $i = 1$):

*DEFINITION 4.22*
*Probabilistic Information Retrieval (PIR) is a CIR* $\Phi = \langle D, \Re \rangle$ *where*
$\Re(q) = \{d | \mu_{\tilde{a}_1}(q, d) \geq \mu_{\tilde{a}_2}(q, d), \mu_{\tilde{a}_1}(q, d) > \alpha_1\}.$

As a noteworthy property it can be easily seen that Similarity Information Retrieval *(SIR)* can be conceived as a special case of *PIR* as follows.

**THEOREM 4.22 [DOMINICH 1999B]**
Let $\Phi = \langle D, \Re \rangle$ be an *PIR*. If $\tilde{a}_1 = \tilde{a}_2$ and $\mu_{\tilde{a}_2}(q, d) = \mu_{\tilde{a}_1}(d, q) \; \forall d, q$, then *PIR* is an *SIR*.

**Proof**. Immediate. ♦

As seen already, two special cases of *PIR* are:

*DEFINITION 4.23*
*Binary Probabilistic Information Retrieval (BPIR) is a PIR with* $S = \{0, 1\}$.

*DEFINITION 4.24*
*Non–Binary Probabilistic Information Retrieval (NBPIR) is a PIR with* $S = [0, 1]$.

In what follows we will demonstrate in two steps that the classical probabilistic *IR* model and *PIR* of Definition 4.22 (which is equivalent to Definition 4.8) are equivalent.

The following result (Theorem 4.23) shows a connection of direct proportionality between the degree $\mu_{\tilde{a}}(q, d)$ of an object document $d$ belonging to a criterion $\tilde{a}$ and conditional probability $P(\tilde{a}|(q, d))$, as well as that Axiom P of Definition 4.8 yields Bayes' Decision Rule (Definition 4.21).

**THEOREM 4.23 [DOMINICH 2000B]**
Given a set $T$, $D$ and $\tilde{a}_i$ as in *CIR*. Let $P$ be a probability measure defined in a $\sigma$–algebra of universe $T$. Let

$$p_{kj}^{(i)} = P(X_k = \mu_{\tilde{o}_j}(t_k)), \; i = 1, 2$$

denote the probability that a random varibale $X_k$ associated to identifier $t_k$ takes on the value $\mu_{\tilde{o}_j}(t_k))$. If

a) $\mu_{\tilde{a}_i}(q, \tilde{o}_j) = \sum_{k=1}^{N} \log (p_{kj}^{(1)} / p_{kj}^{(2)}), \; i = 1, 2;$

b) identifier occurrences are independent;
c) the two criteria are disjoint;

then

1. The degree of compatibility of an object with a criterion is directly proportional to the conditional probability of the criterion given the object, i.e. $\mu_{\tilde{a}_1}(q, \tilde{o}_j) \geq \mu_{\tilde{a}_1}(q, \tilde{o}_s) \Leftrightarrow P(\tilde{a}_1|\tilde{o}_j) \geq P(\tilde{a}_1|\tilde{o}_s)$.
2. $\mu_{\tilde{a}_1}(q, \tilde{o}_j) \geq \mu_{\tilde{a}_2}(q, \tilde{o}_j) \Leftrightarrow P(\tilde{a}_1|\tilde{o}_j) \geq P(\tilde{a}_2|\tilde{o}_j)$.

**Proof.**
1. On can write (condition a):

$$\mu_{\tilde{a}_i}(q, \tilde{o}_j) = \sum_{k=1}^{N} \log (p_{kj}^{(1)} / p_{kj}^{(2)}) = \log \prod_{k=1}^{N} (p_{kj}^{(1)} / p_{kj}^{(2)})$$

Any object $\tilde{o}_j$ can be conceived as a compound event of the simulteneous events with probabilities $p_{kj}^{(i)}$. Thus, a probability $P_i(\tilde{o}_j)$, $\tilde{o}_j \in \tilde{a}_i$, $i = 1, 2$, can be constructed as follows (condition b):

$$P_i(\tilde{o}_j) = \prod_{k=1}^{N} p_{kj}^{(i)}$$

$P_i(\tilde{o}_j)$ can be interpreted as a conditional probability of $\tilde{o}_j$ given relevance $\tilde{a}_1$ or non–relevance $\tilde{a}_2$, and can thus be denoted by $P_1(\tilde{o}_j) = P(\tilde{o}_j|\tilde{a}_1)$ and $P_2(\tilde{o}_j) = P(\tilde{o}_j|\tilde{a}_2)$, respectively. $P_i(\tilde{o}_j)$ is unique for $\tilde{o}_j$ (disjoint criteria, condition c)).
Thus

$$\frac{P_1(\tilde{o}_j)}{P_2(\tilde{o}_j)} = \frac{P(\tilde{o}_j|\tilde{a}_1)}{P(\tilde{o}_j|\tilde{a}_2)} = \frac{\prod_{k=1}^{N} p_{kj}^{(1)}}{\prod_{k=1}^{N} p_{kj}^{(2)}} = \prod_{k=1}^{N} (p_{kj}^{(1)} / p_{kj}^{(2)})$$

Hence

$$\mu_{\tilde{a}_1}(q, \tilde{o}_j) \geq \mu_{\tilde{a}_1}(q, \tilde{o}_s)$$

$$\Leftrightarrow \log \prod_{k=1}^{N} (p_{kj}^{(1)}/p_{kj}^{(2)}) \geq \log \prod_{k=1}^{N} (p_{ks}^{(1)}/p_{ks}^{(2)})$$

$$\Leftrightarrow \prod_{k=1}^{N} (p_{kj}^{(1)}/p_{kj}^{(2)}) \geq \prod_{k=1}^{N} (p_{ks}^{(1)}/p_{ks}^{(2)})$$

$$\Leftrightarrow \frac{P(\tilde{o}_j|\tilde{a}_1)}{P(\tilde{o}_j|\tilde{a}_2)} \geq \frac{P(\tilde{o}_s|\tilde{a}_1)}{P(\tilde{o}_s|\tilde{a}_2)}$$

It follows that:

$$\frac{P(\tilde{o}_j|\tilde{a}_1)}{P(\tilde{o}_j|\tilde{a}_2)} - \frac{P(\tilde{o}_s|\tilde{a}_1)}{P(\tilde{o}_s|\tilde{a}_2)} \geq 0 \Leftrightarrow P(\tilde{o}_j|\tilde{a}_1)P(\tilde{o}_s|\tilde{a}_2) - P(\tilde{o}_s|\tilde{a}_1)P(\tilde{o}_j|\tilde{a}_2) \geq 0$$

By Bayes' Formula we have:

$$P(\tilde{a}_1|\tilde{o}_j) = \frac{P(\tilde{o}_j|\tilde{a}_1)P(\tilde{a}_1)}{P(\tilde{o}_j|\tilde{a}_1)P(\tilde{a}_1) + P(\tilde{o}_j|\tilde{a}_2)P(\tilde{a}_2)}$$

$$P(\tilde{a}_1|\tilde{o}_s) = \frac{P(\tilde{o}_s|\tilde{a}_1)P(\tilde{a}_1)}{P(\tilde{o}_s|\tilde{a}_1)P(\tilde{a}_1) + P(\tilde{o}_s|\tilde{a}_2)P(\tilde{a}_2)}$$

Hence

$$P(\tilde{a}_1|\tilde{o}_j) - P(\tilde{a}_1|\tilde{o}_s)$$

$$= \frac{P(\tilde{o}_j|\tilde{a}_1)P(\tilde{a}_1)P(\tilde{o}_s|\tilde{a}_2)P(\tilde{a}_2) - P(\tilde{o}_s|\tilde{a}_1)P(\tilde{a}_1)P(\tilde{o}_j|\tilde{a}_2)P(\tilde{a}_2)}{[P(\tilde{o}_j|\tilde{a}_1)P(\tilde{a}_1) + P(\tilde{o}_j|\tilde{a}_2)P(\tilde{a}_2)][P(\tilde{o}_s|\tilde{a}_1)P(\tilde{a}_1) + P(\tilde{o}_s|\tilde{a}_2)P(\tilde{a}_2)]}$$

$$= \frac{P(\tilde{a}_1)P(\tilde{a}_2)[P(\tilde{o}_j|\tilde{a}_1)P(\tilde{o}_s|\tilde{a}_2) - P(\tilde{o}_s|\tilde{a}_1)P(\tilde{o}_j|\tilde{a}_2)]}{[P(\tilde{o}_j|\tilde{a}_1)P(\tilde{a}_1) + P(\tilde{o}_j|\tilde{a}_2)P(\tilde{a}_2)][P(\tilde{o}_s|\tilde{a}_1)P(\tilde{a}_1) + P(\tilde{o}_s|\tilde{a}_2)P(\tilde{a}_2)]} \geq 0$$

and thus $\mu_{\tilde{a}_1}(q, \tilde{o}_j) \geq \mu_{\tilde{a}_1}(q, \tilde{o}_s) \Leftrightarrow P(\tilde{a}_1|\tilde{o}_j) \geq P(\tilde{a}_1|\tilde{o}_s)$   (Result 1)

2) Because, by assumption a), we have

$$\mu_{\tilde{a}_i}(q, \tilde{o}_j) = \sum\nolimits_{k=1}^{N} \log(p_{kj}^{(1)}/p_{kj}^{(2)}), \quad i = 1,2,$$

the condition (left hand side of the equivalence 2))

$$\mu_{\tilde{a}_1}(q, \tilde{o}_j) \geq \mu_{\tilde{a}_2}(q, \tilde{o}_j)$$

reduces to an equality:

$$\mu_{\tilde{a}_1}(q, \tilde{o}_j) = \mu_{\tilde{a}_2}(q, \tilde{o}_j)$$

A value or threshold $c$ can always be chosen such that:

$$\mu_{\tilde{a}_1}(q, \tilde{o}_j) = \mu_{\tilde{a}_2}(q, \tilde{o}_j) \geq c$$

This rewrites as follows:

$$\log \prod\nolimits_{k=1}^{N} (p_{kj}^{(1)}/p_{kj}^{(2)}) \geq c$$

$$\Leftrightarrow \prod\nolimits_{k=1}^{N} (p_{kj}^{(1)}/p_{kj}^{(2)}) \geq C = b^c \geq 1 \ (b > 1 \text{ base of logarithm})$$

$$\Leftrightarrow \frac{P(\tilde{o}_j|\tilde{a}_1)}{P(\tilde{o}_j|\tilde{a}_2)} \geq C \geq 1$$

Because $b$ can be chosen such that

$$b^c \geq \frac{P(\tilde{a}_2)}{P(\tilde{a}_1)}$$

it follows that

$$\frac{P(\tilde{o}_j|\tilde{a}_1)}{P(\tilde{o}_j|\tilde{a}_2)} \geq \frac{P(\tilde{a}_2)}{P(\tilde{a}_1)}$$

Using the Bayes' Formula this may be re-written as:

$$\frac{P(\tilde{o}_j|\tilde{a}_1)P(\tilde{a}_1)}{P(\tilde{o}_j)} \geq \frac{P(\tilde{o}_j|\tilde{a}_2)P(\tilde{a}_2)}{P(\tilde{o}_j)}$$

which is equivalent to:

$$P(\tilde{a}_1|\tilde{o}_j) \geq P(\tilde{a}_2|\tilde{o}_j) \qquad \text{(Result 2)} \quad \blacklozenge$$

As a special case of Theorem 4.23, the following holds, too:

**THEOREM 4.24 [YU, MENG AND PARK 1989]**
Given documents $D_i$, $i = 1, ..., M$, as frequency vectors

$$D_i = (d_{ik}|\ k = 1, ..., N)$$

and a dot product similarity function

$$f(X, Y) = \sum_{k=1}^{N} x_k y_k$$

Let

$$D'_i = (d'_{ik}|\ k = 1, ..., N)$$

be new frequency vectors where

$$d'_{ik} = \log (P(d_{ik}|R) / P(d_{ik}|I))$$

where $P(d_{ik}|.)$ is the probability that a relevant/irrelevant document has $d_{ik}$ occurrences of the $k$th term. Then

$$P(D_i|R) > P(D_j|I) \Leftrightarrow f(Q, D_i) > f(Q, D_j)$$

where $Q$ denotes a query with all weigths equal to 1 for terms present and 0 otherwise.

**Proof.** Immediate; $\mu_{\tilde{o}_j}(t_k) =$ number of occurrences of term $t_k$ in $\tilde{o}_j$. $\blacklozenge$

Yu, Meng and Park [1989], who first proved — following a different line — this result (Theorem 4.24), consider it as an *optimal way* in which to retrieve documents.

It can now be shown that the traditional probabilistic *IR* (see part 1.3 or Definition 4.21) and *PIR* (Definition 4.8), as a special case of *CIR*, are equivalent.

## THEOREM 4.25 [DOMINICH 2000B]

The two *PIR*s are equivalent.

**Proof.** Property P2 becomes ($C = 2$, two criteria):

$$\mathfrak{R}(q) = \{\delta \mid \mu_{\tilde{a}_i}(q, \delta) = \max_{k=1, \ldots, C} \mu_{\tilde{a}_k}(q, \delta)\} \cap a_{\alpha_i}$$

$$= \{\delta \mid \mu_{\tilde{a}_i}(q, \delta) \geq \mu_{\tilde{a}_j}(q, \delta), \ j = i + (-1)^{i+1}, \ \mu_{\tilde{a}_i}(q, \delta) > \alpha_i\}.$$

Based on Theorem 4.23, this rewrites as follows:

$$\mathfrak{R}(q) = \{\delta \mid P(\tilde{a}_i \mid \delta) \geq P(\tilde{a}_j \mid \delta), \ j = i + (-1)^{i+1}, \ P(\tilde{a}_i \mid \delta) > \alpha_i\}$$

which is exactly the retrieval condition in *PIR* (*i* means 'relevant'). Property P1 ensures that reflexivity holds, too. ◆

In order to apply the *PIR* in practice, one can proceed as follows. Let

$$T = \{t_1, \ldots, t_k, \ldots, t_N\} \tag{4.30}$$

be a set of identifiers and

$$D = \{\delta_j \mid \delta_j = \{(t_k, \mu_{\delta_j}(t_k)) \mid t_k \in T, \ k = 1, \ldots, N\}, \ j = 1, \ldots, M\} \tag{4.31}$$

a set of objects.

Let $\mu_{\delta_j}(t_k) = n_{kj}$ be the number of occurrences of identifier $t_k$ in object $\delta_j$; $n_{kj} \in \{0, 1\}$ for the *BPIR* with $n_{kj} = 1$ denoting the presence and $n_{kj} = 0$ the absence of the identifier, and $n_{kj} \in [0, 1]$ for the *NBPIR* with $n_{kj}$ denoting e.g. a (normalised) frequency of the identifier ($n_{kj}$ = number of occurrences of $t_k$ in $\delta_j$ divided by total number of occurrences of identifiers in $\delta_j$).

Let

$$\tilde{a}_i = \{((q, \delta_j), \mu_{\tilde{a}_i}(q, \delta_j)) \mid \delta_j \in D, \ j = 1, \ldots, M\} \tag{4.32}$$

be two criteria, where $q \in D$, $i = 1$ (relevant), 2 (non–relevant) two criteria.

Let $P$ be a probability measure defined in a $\sigma$–algebra of universe $T$. Let $p_{kj}^{(i)}$ denote the probability that a random variable $X_k$ associated to term $t_k$ takes on the value $\mu_{\delta_j}(t_k)$ in $\tilde{a}_i$, $p_{kj}^{(i)} = P(X_k = \mu_{\delta_j}(t_k))$, $i = 1, 2$.

We have seen that in order for an object $\delta_j$ to be considered as relevant, i.e. to be retrieved in response to a query $q$, the following condition should be fulfilled:

$$P(\tilde{a}_1|\delta_j) \geq P(\tilde{a}_2|\delta_j) \tag{4.33}$$

Using Bayes' Theorem

$$\frac{P(\delta_j|\tilde{a}_i)P(\tilde{a}_i)}{P(\delta_j)}, \quad i = 1, 2$$

the retrieval condition re–writes as follows:

$$\sum_{k=1}^{N} \log (p_{kj}^{(1)} / p_{kj}^{(2)}) \geq c \quad \text{where } c = -P(\tilde{a}_1)/P(\tilde{a}_2) \tag{4.34}$$

The left hand side of 4.34 is called a *discriminant function*, and is used as a retrieval function in practice. We can also write:

$$\sum_{k=1}^{N} \log p_{kj}^{(1)} \geq \sum_{k=1}^{N} \log p_{kj}^{(2)} + c \tag{4.35}$$

The probabilities $p_{kj}^{(1)}$, $p_{kj}^{(2)}$ are estimated as follows: The user is presented a subset

$$D' = \{d'_1, ..., d'_m\} \subseteq D \tag{4.36}$$

of documents and divides $D'$ into two disjoint subsets $D'_1$ and $D'_2$ (this is referred to as *relevance feedback*):

$$D'_1 \cup D'_2 = D', D'_1 \cap D'_2 = \varnothing \tag{4.37}$$

Let

$$D'_1 = \{d'_{11}, ..., d'_{1k}, ..., d'_{1K}\} \tag{4.38}$$

be the subset of documents judges as being relevant, and

$$D'_2 = \{d'_{21}, ..., d'_{2s}, ..., d'_{2S}\} \qquad (4.39)$$

be the subset of documents judged as being non–relevant. A probability

$$p_{kj}^{(i)} = P(\mu_{\delta j}(t_k)|i), \; i = 1, 2 \qquad (4.40)$$

is computed, and thus explicit forms for the discriminant function are obtained.

A practical method for such a computation is presented below. The following table is constructed for each identifier $t_k$:

| 0 | 1 | 2 | $n_k$ | |
|---|---|---|---|---|
| $c_0$ | $c_1$ | $c_2$ | $c_{nk}$ | for $D'_1$ |
| $e_0$ | $e_1$ | $e_2$ | $e_{nk}$ | for $D'_2$ |

where $n_k$ denotes the number of occurrences ($n_k = 0, 1, 2, ...$), $c_{nk}$ ($e_{nk}$) denotes the number of relevant (non–relevant) documents having $n_k$ ocurrences of $t_k$. Let us denote probability in 4.40 by $p(i)_{kz}$, and it is computed according to the formula:

$$p(i)_{kz} = \frac{n_k}{Z} \qquad Z = K \text{ for } D'_1 \text{ and } S \text{ for } D'_2 \qquad (4.41)$$

Once the probabilities $p_{kj}^{(i)}$ have been calculated, new weights $w_{ij}$ are computed for objects using the formula $w_{ij} = \log (p(i)_{kK} / p(i)_{kS})$. Then the dot product similarity measure is calculated for every pair (documents_weights_vector, query_weights_vector). The documents are rankordered and a cut off value can also be used.

See Appendix 2 for a complex example and complete procedures (in *MathCAD 8 Plus Professional*).

## 4.1    Conclusions

As opposed to the *SIR*, where the retrieval function is commutative, the retrieval function $P(\tilde{a}|\tilde{o})$, on which the *PIR* is based, is not commutative.

The SIR is a special case of *PIR* provided that commutativity is imposed on the latter. Thus, in a sense, purely formally, the presence or absence of commutativity decides which of the two we have.

## 4.2    Bibliographical Remarks

The main objective of an information retrieval system is to identify those documents which are relevant to a particular user. Albeit it seems impossible to design a system that will strictly predict the relevant documents, a theory of designing a system in which retrieval is based on estimating a degree or probability of relevance has taken shape after undergoing several stages. This theory has led to a basic mathematical model of *IR* called Probabilistic Information Retrieval (*PIR*).

Maron and Kuhns [1960] formulated a basic mathematical model of Probabilistic Information Retrieval (*PIR*) called Model 1. This is usually stated as follows: Given a document *D* and a term *T* estimate the probability that a user would select *T* if he/she would want to find *D*. The estimations are owed to a human indexer.

Robertson and Sparck Jones [1976] formulated another basic mathematical model of *PIR*, called Model 2, which is usually stated as follows: the user is shown a series of documents and judges which are relevant and which are not. Based on this feed back from the user, an estimate of a probability of relevance/non–relevance for each term is calculated, according to its distribution in the relevant/non–relevant documents. This probability would then be used to estimate the probability of relevance/non–relevance of an arbitrary document.

Robertson, Maron and Cooper [1982] unified Model 1 and Model 2 in one model called Model 3: Both estimations by the two human factors, indexer and user, are present and made use of.

A basic mathematical model for *PIR* has also taken shape and developed as a result of the work of Van Rijsbergen [1977, 1979], Yu and Salton [1976], Salton and McGill [1983], Yu, Meng and Park [1989], Wong and Yao [1990].

See also part 1.3.1.

## 5.    INTERACTION INFORMATION RETRIEVAL

In a *CIR* object documents are typically not effectively interconnected, and even if they were so the object query does not affect the interconnections. Further, retrieval is based on calculating a numerical value (similarity, probability) for pairs of objects (namely object query and object documents). This numerical value may be conceived as an expression of a degree of connection (similarity, likeness, probability) between objects. These 'pseudo–links' are static (i.e., the presence of a new object, e.g., object–

query, does not influence nor modify the existing pseudo–links between the objects).

In *Interaction Information Retrieval* (*IIR = I²R*, see also Chapter 3), the connections between objects are made explicit, and are used. There are multiple connections between object–documents. Each time a new object is added it is interconnected (new links) to the already existing and interconnected objects and some of the old links may change. The new links and the changed old links represent a real interaction between the new object and the other objects. Thus the object query is also interconnected with the other objects (before being answered): it becomes a member of the already interconnected system of object documents, causing some of the interconnections to change (interaction). Retrieval is defined as a result of this interaction, and depending on how this is defined, one gets:

1. Associative *I²R* (*AI²R*). An activation originates at the object query and spreads step by step, from winner to winner. At a certain point reverberative circles (loops) are formed, when activation should be halted or else it will loop forever. The reverberative circles are interpreted as local memories recalled by (or associated to) the object–query.
2. *CIR* as a special case of *I²R*. Only the first step of activation spreading is performed, from object–query to the first winner(s).

## 5.1     D–Net

Given a set

$$D = \{d_1, d_2,..., d_i,..., d_M\} \tag{4.42}$$

of elements called *documents*, $M > 1$. An Artificial Neural Network (*ANN*)

$$D\text{–}Net = \langle \aleph, W, L \rangle \tag{4.43}$$

is associated to $D$ as follows. The set

$$\aleph = \{v_i|\ v_i \text{artificial neuron assigned to } d_i, i = 1, 2, ..., M\} \tag{4.44}$$

denote a set of artificial neurons. Further let

$$L: \aleph \times \aleph \to \mathbf{R}_+^{K_{ij}}, L(v_i, v_j) = \mathbf{w}_{ij} = (w_{ij}^1 ,..., w_{ij}^k ,..., w_{ij}^{K_{ij}}) \in \mathbf{R}_+^{K_{ij}}$$
$$\tag{4.45}$$

denotes *connection strengths* or *weights*, $i, j = 1, 2,..., M$, and

$$W = \{\mathbf{w}_{ij} | \ i, j = 1, 2,..., M, i \neq j\} \tag{4.46}$$

denotes a set of weights. As the connections are meant to appear between documents when they share common meaning, it is natural to assume that the following conditions hold (to express that sharing common meaning is not 'one–way'):

$$\mathbf{w}_{ij} \neq 0 \Rightarrow \mathbf{w}_{ji} \neq 0, \ \forall i, j; \ 0 \le \mathbf{w}_{ij}^{k} \le 1, \ \forall i, j, k \tag{4.47}$$

An *activation spreading* takes place in a *D–Net*. The *state* (or *activity level*) of $v_i$ is denoted by $z_i$. Activation spreads according to a winner–take–all (WTA) strategy [Feldman and Ballard 1982], from an element to another (different) element.

In a *D–Net*, considering a new object *o* (document or query) means the set union $D' = \aleph \cup \{v_o\}$, and, at the same time, $L$ of Definition 4.45 is applied for every neuron pair again. Thus the following concept is introduced:

*DEFINITION 4.25*
*Feeding* a new artificial neuron $v$ into a *D–Net* $= \langle \aleph, W, L \rangle$ means obtaining a new $D'$–*Net* $= \langle \aleph', W', L' \rangle$ where

$$\aleph' = \aleph \cup \{v\}, L' : \aleph' \times \aleph' \rightarrow \mathbf{R}_+^{K_v} \tag{4.48}$$

$W'$ contains more weights than $W$ hence $|W'| > |W|$, $W' \setminus W \neq \emptyset$. One can distinguish the following three cases:

1. $W'$ contains the new weights for the fed $v$ and all of the old weights but unchanged,
2. $W'$ contains the new weights for the fed $v$ and all of the old weights but changed,
3. $W'$ contains the new weights for the fed $v$ and both changed and unchanged old weights.

## 5.2    Interaction Information Retrieval

Interaction Information Retrieval ($I^2R$) is now defined as follows:

*DEFINITION 4.26* [DOMINICH 1999B]

Let $\langle \aleph, W, L \rangle$ be a $D$–Net and $\langle \aleph', W', L' \rangle$ an associated fed $D'$–Net.. *Interaction Information Retrieval* ($IIR=I^2R$) on $D'$ is an *IRF* $\Phi = \langle D, \mathfrak{R} \rangle$ where $\mathfrak{R}(q) = \{ d_i \mid v_i \text{ is a winner}\}$.

Depending on how far a WTA strategy–based activation is permitted, one gets an *Associative* $I^2R$ ($AI^2R$) or — as a special case — a *CIR* (i.e., $I^2R$ behaves like an *CIR*).

## 5.2.1 Associative Interaction Model

Retrieval is defined as winners in an activation originating at the fed object. It is shown that a long enough activation yields reverberative circles (loops) which act as local memories triggered by (or associated to) the fed object.

For technical reasons, it is convenient to introduce the following concepts first.

*DEFINITION 4.27*

A *reverberative circle* $\varsigma$ is a sequence $\varsigma = v',...., v^p,...., v^V$ of artificial neurons where: $v^p$ denotes the $p$th neuron, $v' = v^V$ and $v^p$ is a winner, i.e. is the most active of all elements succeeding its predecessor, i.e. $v^p$ such that

$$z_p = \max_j \{z_j \colon \mathbf{w}_{p-1,j} \neq 0, \; p-1 \neq j\} \tag{4.49}$$

*DEFINITION 4.28*

An element $v$ is an *isolated element* of $D'$–Net if all incoming and outgoing links have weights equal to zero, i.e. $\mathbf{w}_{vj} = \mathbf{w}_{jv} = 0 \; \forall j$.

*DEFINITION 4.29*

An element $v$ *recalls* a reverberative circle $\varsigma$ if $\varsigma$ is formed owing to an activation originating at $v$.

Note

A reverberative circle does not necessarily contain the element it originates from.

The question of whether there are reverberative circles at all is answered by the following:

## THEOREM 4.26 [DOMINICH 1999C]

There exists at least one reverberative circle $\varsigma$ in a $D'$–Net recalled by a non–isolated element $v$.

**Proof.** Let an activation start from $v$ and passing on to another element $v'$.
There are only two possibilities:
a) either the activation goes back to $v$ (in which case $v$ and $v'$ form a
   reverberative circle),
b) or it spreads on to another element $v''$, and so on until it either reaches $v$
   (where it initially started from in which case we have a reverberative
   circle), or some other element previouly already reached in which case
   we also have a reverberative circle ($D'$ finite). ♦

It is immediate that:

**THEOREM 4.27**
An isolated element $v$ recalls no reverberative circles.

**Proof.** Because $w_{vj} = 0$ for all $j$ no activation starts from $v$. ♦

Associative Interaction Information Retrieval ($AI^2R$) is now defined as
follows:

*DEFINITION 4.30* [DOMINICH 1999B]
Let $\langle \aleph, W, L \rangle$ be a $D$–Net and $\langle \aleph', W', L' \rangle$ an associated fed $D'$–Net..
*Associative Interaction Information Retrieval ($AI^2R$) on $D'$ is an IRF* $\Phi = \langle D, \mathfrak{R} \rangle$ where $\mathfrak{R}(q) = \{ d_i \mid v_i \in \varsigma, \varsigma \text{ recalled by } v_q \}$.

It can be easily seen that:

**THEOREM 4.28**
$v_q$ is isolated $\Leftrightarrow \mathfrak{R}(q) = \varnothing$.

**Proof.** If $v_q$ is isolated then it recalls no reverberative circle, hence $\mathfrak{R}(q) = \varnothing$. Assuming now the opposite, i.e. $\mathfrak{R}(q) = \varnothing$, $v_q$ must be isolated or else it
would recall at least one reverberative circle. ♦

Theorem 4.28 makes it possible to define a no–hit as follows:

*DEFINITION 4.31*
A no–hit is an isolated element.

Taking into account $W$ and $W'$ (see above), it is possible to make a
distinction between a real and a pseudo–interaction, as follows:

DEFINITION 4.32

Let $\langle \aleph, W, L \rangle$ be a *D–Net* and $\langle \aleph', W', L' \rangle$ be the fed *D'–Net*. The set difference $W' \setminus W$ is *interaction* $I$ between $v$ and *D–Net*, $I = W' \setminus W$. When only a) occurs, interaction is called a *pseudo–interaction*: $I = W' \setminus W = \{\mathbf{w'}_{vj}, \mathbf{w'}_{jv}, \forall j\}$. When b) or c) or both occur, interaction is called a *real interaction*: $I = W' \setminus W = \{\mathbf{w'}_{vj}, \mathbf{w'}_{jv}, \exists i, j: \mathbf{w'}_{ij} \neq \mathbf{w}_{ij}\}$.

Two measures for interaction can be defined as follows: a size to indicate 'how much' of $\aleph$ is affected, and an intensity to indicate 'how strong' the interaction is.

DEFINITION 4.33

The *size S* of interaction $I$ is equal to $S = |W' \setminus W| = |I|$.

DEFINITION 4.34

The *intensity* $\iota$ of interaction $I$ is equal to $\iota = \sum_I w^p_{ij}$.

It is now shown that the abstract mathematical structure characterising retrieval is as follows:

**THEOREM 4.29 [DOMINICH 2000A]**

Retrieval in $AI^2R$ means defining a matroid.

**Proof.** $\langle \aleph', W', L' \rangle$ can be assigned a complete, directed, weighted multigraph $G$ as follows:

a) each artificial neuron $v_i$ is assigned a vertex $v_i$,

b) there are two oppositely directed edges (opposite arcs), $e_{ij}$ and $e_{ji}$, between every pair of vertices $v_i$ and $v_j$ ($i \neq j$) having weights $u_{ij}$ and $u_{ji}$ respectively, where

$$u_{ij} = \sum_{k=1}^{K_{ij}} w^k_{ij}$$

and

$$u_{ji} = \sum_{k=1}^{K_{ji}} w^k_{ji}.$$

Any reverberative circle $\varsigma$ corresponds to a circle $C$ in graph $G$. Let

$$N = \{v_\alpha | \alpha = 1, 2, ..., A\}$$

denote the artificial neurons traversed before $\varsigma$ is recalled by $v_q$. Then $N$ corresponds to a path

$$P = \{v_\alpha | \alpha = 1, 2, ..., A\}.$$

This means that retrieval defines a connected subgraph $H$ with circles and cutpoints. Hence a block cutpoint graph $T$ can be assigned to subgraph $H$, and a matroid is generated (cycle matroid). ♦

Thus, for a practically implemented $AI^2R$, the ideal or optimal retrieval case can be defined as follows:

*DEFINITION 4.35*
An implemented $AI^2R$ is *optimal* if $\mathfrak{R}(q)$ is the associated cycle matroid.

As another important result as to practice, it is now shown that the complexity of the computation of numerical values involved is as follows:

**THEOREM 4.30 [DOMINICH 1999C]**
The complexity of computations is polynomial.

**Proof.** There are $2(n_i + n_j)$ number of weights between every pair $(o_i, o_j)$ of which there are $^M C_2$, hence $2(n_i + n_j) \, ^M C_2 = O(NM^2)$ is an upper bound for weights computation, where $N = \max_{i,j}(n_i, n_j)$, i.e., the largest of object lengths. The computation of the sums of the weights between a given object $o_i$ and all the other objects $o_j$, of which there are $M - 1$ is $(n_i + n_j)(M - 1)$, and thus an upper bound for the computation of all sums is $(n_i + n_j)(M - 1)^2 = O(NM^2)$ because $i$ can vary, too, at most $M - 1$ times. An upper bound to find the strongest connection from a given object to all the others is $O(M)$ (finding the maximum from a sequence of numbers), and thus an upper bound for the selection of all strongest connections in the entire network is $O(M)(M - 1) = O(M^2)$. Hence an overall upper bound for the weights computation is $O(NM^2) + O(NM^2) + O(M^2) = O(NM^2) = O(K^3)$, where $K = \max (N, M)$. ♦

   Note
Finding a reverberative circle seems to be equivalent to the problem: "Does an undirected graph contain a simple cycle of at least $k$ nodes?" which is known to be **NP-Complete**. Because drawing the corresponding graph after retrieval has taken place is very similar to the above problem (although not all circles may be simple) it can be hypothesised that $AI^2R$ is NP–complete too, but this needs a proof, and thus is an open problem.

Activity level $z_i$ can be expressed using the General Network Equation:

$$\frac{dz_i(t)}{dt} = I_i(t) - z_i(t) + \sum_j T_{ij} f_j(z_j(t)) \tag{4.50}$$

where $z_i$ denotes the *activity level* of element $v_i$, $I_i(t)$ denotes the external input to $v_i$, $t$ denotes time, $f_j(z_j(t))$ denotes the influence of $v_j$'s activity level on $z_i$ and $T_{ij}$ denotes connection strengths between $v_i$ and $v_j$. Because there is no external input we take $I_i(t) = 0$. $T_{ij}$ corresponds to $w_{ij}$, and it can be taken as, e.g., the sum of all weights. The term $f_j(z_j(t))$ is taken as unity as it is assumed that all elements have equal influence. Thus the network equation becomes:

$$\frac{dz_i(t)}{dt} = -z_i + \sum_j \sum_k w_{ji}^k \tag{4.51}$$

Let $v_q$ denote an artificial neuron under focus. An activation is started and spread from $v_q$ at time $t_0$ by clamping its state to 1, i.e., $z_q(t_0) = 1$, and $z_i(t_0) = 0$, $\forall i \neq q$. At $t_1 > t_0$, the maximum activity *max* $z_i$ is to be found from among all $i$, $z_i$ being influenced by $z_q$, i.e., $j = q$. $z_i$ is given by the solution of the following Cauchy problem:

$$\frac{dz_i(t)}{dt} = -z_i + \sum_k w_{ji}^k, \qquad z_i(t_0) = 0, \ \ j = q \tag{4.52}$$

Denoting the sum by $s_i$, the solution is

$$z_i(t) = s_i \, e^{-t} (e^t - 1) \tag{4.53}$$

Assuming now that $v$ has the maximum activity $z$ of all its 'competitors' $v_i$, we have:

$$z(t) \geq z_i(t) \ \Leftrightarrow \ (s - s_i) \, e^{-t} (e^t - 1) \geq 0 \tag{4.54}$$

Because $e^{-t}(e^t - 1)$ is positive it follows that $s \geq s_i$. In other words, maximum activity is equivalent to maximum weights sum:

$$\sum_k w_{ji}^k \tag{4.55}$$

EXAMPLE
See Appendix 3 for a complex example along with complete procedures )in MathCAD 8 Plus Professional). Alternatevily, see Example 3.5 (Chapter 3).

## 5.2.2    Classical Information Retrieval: Pseudo–interaction

It can be shown that $CIR$ is a special interpretation of $I^2R$. The idea is that only the first step of activation spreading is performed, from object–query to the first winner(s).

This is expressed in the following:

**THEOREM 4.31 [DOMINICH 2000A]**
Given an $I^2R$, an arbitrary fixed $k$ and a threshold value $\alpha_k$. If

i)     $K_{ij} = C = $ constant $\geq 1$, $\forall$ $i,j$

ii)    $L(v_i, v_i) = 1$ (reflexivity)

iii)   $z_j \leftarrow$ *if* $w_{qj}^k = \max_v w_{qj}^v \geq \alpha_k$ *then 'winner' otherwise* 0, i.e., the winner is that element $j$ whose $k$th input exceeds some threshold value, and is the highest of all inputs,

iv)   after the first step (i.e., after starting the activation spreading from $v_q$), stop activation,

then $I^2R$ behaves like an $CIR$, i.e. $CIR$ is speacial case of $I^2R$.

**Proof.**
It is shown that an $I^2R$ with conditions i)—iv) satisfies properties P1 and P2 for $CIR$.

Condition i) $K_{ij} = C = $ constant, $\forall$ $i,j$, is viewed as denoting the number $C$ of criteria.

$L$ is viewed as corresponding to membership function $\mu$ with $w_{ij}^k$ corresponding to $\mu_{\tilde{a}_k}(\tilde{o}_i, \tilde{o}_j)$.

Condition ii), $L(v_i, v_i) = 1$ (reflexivity), corresponds to property P1 for $CIR$.

$\mathfrak{R}(q)$ in $I^2R$ is equal to $\mathfrak{R}(q) = \{d_j | v_j$ is a winner$\}$ which — according to condition iii) — re–writes as

$$\mathfrak{R}(q) = \{d_j | v_j, w_{qj}^k = \max_v w_{qj}^v \geq \alpha_k\}$$

which is identical to property P2 of $CIR$, i.e. ,

$$\mathfrak{R}^k(q) = \{\tilde{o}_j | \mu_{\tilde{a}_j}(q, \tilde{o}_j) = \max_{\nu=1 \ c} \mu_{\tilde{a}_\nu}(q, \tilde{o}_j) \} \cap a_{\alpha_k}. \ \blacklozenge$$

Notes

It is very important to note an underlying difference between *CIR* and $I^2R$. The interconnections in the *D*–Net are indifferent from the point of view of Theorem 4.31. This is perhaps the main difference between a 'true'*CIR* (vector, probabilistic) and a 'true' $I^2R$ ($AI^2R$): an $AI^2R$ does make effective changes and use of interconnections, while a 'true' *CIR* does not take interconnections into account at all even if they existed. Or, in other words, in a 'true' $I^2R$ a real interaction takes place, while in a 'true' *CIR* only a pseudo–interaction takes place, i.e., the object–query does not modify the structure of object documents before being answered.

Theorem 4.31 could also be used as a definition for 'true' *CIR*.

It is possible to define an equivalence relation between neurons, as follows:

*DEFINITION*

Two non–isolated elements, $v_1$ and $v_2$, are *equivalent* if they both recall the same reverberative circles, and are *partially equivalent* if they recall at least one identical reverberative circle.

It can be easily seen that this equivalence is an equivalence relation, whilst the partial equivalence is similar to the congruence relation in number theory. For equivalent neurons, the respective circles act as 'attractors'.

# 6. BOOLEAN INFORMATION RETRIEVAL

In this part, mathematical properties of the Boolean model are revealed, and a new light is shed upon this model.

## 6.1 Classical Boolean Model

Let us recall (see part 1.1.) the definition of the classical Boolean model of *IR*. Given a set

$$T = \{t_1, t_2, ..., t_j, ..., t_m\} \tag{4.56}$$

of elements called *index terms* (e.g. words or expressions describing or characterising documents such as keywords given for a journal article), and a set

$$D = \{D_1, ..., D_i, ..., D_n\}, D_i \in \wp(T) \tag{4.57}$$

of elements called *documents* (or representations). Let $Q$ be a Boolean expression, called a *query*, in a conjunctive normal form:

$$Q = \bigwedge_{k \in K} ( \bigvee_{j \in J} \theta_j), \theta_j \in \{t_j, \neg t_j\} \tag{4.58}$$

where $\neg t_j \in D_i$ should be read as $t_j \notin D_i$. Equivalently, $Q$ can be given in disjunctive normal form, too. The set $\Re(Q)$ of retrieved documents in response to query $Q$ is as follows:

$$\Re(Q) = \bigcap_{k \in K} ( \bigcup_{j \in J} S_j) \tag{4.59}$$

$$S_j = \{D_i | \theta_j \in D_i\} \tag{4.60}$$

The Boolean model of $IR$ has a great practical importance in that the majority of commercial $IR$ systems, especially database applications, are based on it.

Usually, perhaps mainly by virtue of tradition, the Boolean model of $IR$ is considered to be an elementary and distinct model of $IR$.

In principle, however, the Boolean model may be conceived in a somewhat more special way, as we will see.

First, let us observe that the document selecting condition $\theta_j \in D_i$ is equivalent to either $t_j \in D_i$ or $t_j \notin D_i$. Of course, it is sufficcient to consider just the first case, $t_j \in D_i$, as the other is logically equivalent to it. This condition can be written in several equivalent forms, for example:

$$t_j \in D_i \Leftrightarrow \{t_j\} \cap D_i \neq \varnothing \tag{4.61}$$

$$t_j \in D_i \Leftrightarrow |\{t_j\} \cap D_i| \neq 0 \tag{4.62}$$

$$t_j \in D_i \Leftrightarrow (\mathbf{t}_j, \mathbf{D}_i) \neq 0 \tag{4.63}$$

$$t_j \in D_i \Leftrightarrow \text{Cosine}(\mathbf{t}_j, \mathbf{D}_i) \neq 0 \tag{4.64}$$

where $\mathbf{t}_j$ is the binary vector representing $t_j$ whereas $\mathbf{D}_i$ is the binary vector representing $D_i$.

Thus, a similarity $\sigma$ can be defined e.g. using the Cosine:

$$\sigma(t_j, D_i) = 1 \text{ if } \text{cosine}(\mathbf{t}_j, \mathbf{D}_i) \neq 0, \text{ and } 0 \text{ otherwise} \tag{4.65}$$

As seen already, this $\sigma(t_j, D_i)$ is a similarity (it is normalised, symmetric and reflexive).

In this way, the document selecting condition is equivalent to a $SIR_j$ over $D$ yielding $S_j$. Thus, the Boolean $IR$ can be conceived as being defined over $SIR_j, j \in J$.

This yields the following more general definition of the Boolean model of $IR$.

Let $\langle D, \mathfrak{R} \rangle$ be an $IRF$, e.g. $I^2R$, i.e. $AI^2R$ or $CIR$ ($SIR$ or $PIR$), and let $Q = \varepsilon(t_j)$ be a Boolean expression over terms $t_j, j \in J$, in a normal form.

Then, the Boolean model of $IR$ can be defined as follows:

*DEFINITION 4.36*
Given an $IRF$ $\langle D, \mathfrak{R} \rangle$, and let a query $Q = \varepsilon(t_j)$ be a Boolean expression over $t_j$ in a normal form. *Boolean IR (BIR)* over $\langle D, \mathfrak{R} \rangle$ is an $IRF$ $\langle D, \beta \rangle$ where $\beta = \lambda(\mathfrak{R}(t_j))$, where $\lambda$ denotes a logic (e.g., Boolean, Fuzzy) counterpart for $\varepsilon$.

In words, for every query term $t_j$, a set $\mathfrak{R}(t_j)$ of documents is retrieved first. For this retrieval, any other $IR$ model can be used: vector, probabilistic, interaction, logic, etc.. Then these retrieved sets $\mathfrak{R}(t_j)$ are 'combined' using the corresponding logical operators from $\lambda$ (such as, for example, $\wedge$ for $\cap$ and $\vee$ for $\cup$).

It can now be easily seen that the following connection between the Booelan model of $IR$ and mathematical structures can be shown:

**THEOREM 4.32**
Taking all possible $BIRs$ $\langle D, \beta \rangle$ over $D$ is equivalent to a similarity space or topology or recursion or matroid.

**Proof.** The mathematical structure with which $BIR$ is equivalent depends on the particular $IR$ model within which $\mathfrak{R}(t_j)$ is obtained. ◆

Notes
In principle, it would be possible to use different retrievals for different terms. For example, $S_1$ could be obtained using a similarity $IR$, whilst $S_2$ using a probabilistic $IR$. Hence, the Boolean $IR$ model might be conceived in an even more general and complex way.
Thus, the Boolean model is 'more' than any other model in that a final answer to a query is obtained by aggregating the partial answers for individual query terms. It is, at the same time, 'less' than any other model in that these are needed to get partial answers for individual query terms.

As a consequence, an important particular case of Theorem 4.32, for $SIRs$, is the following

**THEOREM 4.33 [EGGHE AND ROUSSEAU 1998]**
The set of all possible Boolean retrievals, based on elementary queries, using
threshold, is equivalent to the retrieval topology; using close matches, is
equivalent to the Euclidean topology.

**Proof**. Immediate. ♦

## 6.2     Weighted Boolean Model

Terms — in document or query or both — are associated *weights*, and
retrieval takes them into account.

A *weight* $w_{ij} \in [0, 1]$ of a term $t_j$ can be an expression of how important
that term is relative to document $D_i$.

For example, the weight $w_{ij}$ can be the number of occurrences of term $t_j$ in
document $D_i$ divided by the total number of terms in $D_i$ (see Chapter 3).
Thus, any document $D_i$ can be associated a weights vector $\mathbf{w}^i$ as follows:

$$\mathbf{w}^i = (w_{i1}, ..., w_{ij}, ..., w_{im}) \tag{4.66}$$

Other definitions of weights have also been proposed [see, e.g., van
Rijsbergen, 1979; Salton, 1983, or see part 2, the Vector *IR*].

Note
In principle, $t_j$ need not effectively occur in $D_i$.

More recently, other weights have been suggested taking into account the
internal structure of a document, e.g. the structure of a HyperText Markup
Language (HTML) document on the World Wide Web (WWW) [see e.g.
Bordogan and Pasi, 1995].

In the Weighted *BIR*, the query terms are associated weights, too. They
express a relative importance of search terms. Thus, formally, a query $Q$ is
as follows:

$$Q = \underset{k \in K}{\wedge} ( \underset{j \in J}{\vee} (\theta_j, v_j)), \; \theta_j \in \{t_j, \neg tj\} \tag{4.67}$$

Because any query can be transformed in a (disjunctive or conjunctive)
normal form, it is enough, formally, to consider two basic cases:

1. AND–ed terms query:

$$Q_\wedge = \langle v_1, t_1 \rangle \wedge \langle v_2, t_2 \rangle \tag{4.68}$$

2. OR–ed terms query:

$$Q_v = \langle v_1, t_1 \rangle \vee \langle v_2, t_2 \rangle \tag{4.69}$$

For retrieval purposes the document selection condition is based on an *evaluation measures* g which is an expression of the degree to which a document can be selected (referred to as Retrieval Status Value = *RSV*). More common evaluation measures are:

[Kraft and Buell, 1983]:

$$g_i(w_{ij}, v_j) = A \cdot \frac{w_{ij}}{v_j}, \quad w_{ij} < v_j \tag{4.70}$$

$$g_i(w_{ij}, v_j) = A + B \cdot \frac{w_{ij} - v_j}{1 - v_j}, \quad w_{ij} \geq v_j \tag{4.71}$$

[Bordogna and Pasi, 1991}:

$$g_i(w_{ij}, v_j) = e^{K \cdot (w_{ij} - v_j)^2} \tag{4.72}$$

[Kraft et al., 1995]:

$$g_i(w_{ij}, v_j) = A \cdot e^{K \cdot (w_{ij} - v_j)^2}, \quad w_{ij} < v_j \tag{4.73}$$

$$g_i(w_{ij}, v_j) = A + B \cdot \frac{w_{ij} - v_j}{1 - v_j}, \quad w_{ij} \geq v_j \tag{4.74}$$

where $A = (1 + v_j)/2$ and $B = (1 - v_j^2)/4$. Just as in the fuzzy *IR* model, disjunction (logical OR) is defined as the 'min', conjunction (logical AND) as the 'max', whereas negation (logical NOT) as 'one minus'. Alternatively, other aggregation operators can also be used for disjunction ([see, e.g., Kraft, et al. 1998]:

$$\min(1, g_i + g_k) \tag{4.75}$$

$$g_i + g_k - g_i g_k \qquad\qquad\qquad\qquad (4.76)$$

and for conjunction:

$$\max(0, g_i + g_k - 1) \qquad\qquad\qquad\qquad (4.77)$$

$$g_i g_k \qquad\qquad\qquad\qquad (4.78)$$

A threshold (cut–off) value can then be used to decide on whether a specific document should effectively be retrieved or not.

The above $g_i$s contain, as a special case, the document selecting condition for the classical Boolean $IR$ ($v_j = 1$, $A = 1$, $B = 0$, $K \to -\infty$, $g_i \in \{0, 1\}$). None of the $g_i$s is a similarity, because, for example, 4.70 and 4.74 are not commutative, and 4.72 cannot be always normalised (e.g. $K \to -\infty$ and $w_{ij} = v_j$). Thus in a — realistically — weighted Boolean $IR$, i.e., when the weights are different from 1 and 0 (they are not 'just' Boolean values), the document selection condition is not equivalent to an $SIR$. As it is also different from the probabilistic model (the $g_i$s) it can be said that a weighted Boolean $IR$ model is a distinct model of $IR$ (we have seen that the traditional Boolean $IR$ model, albeit a particular case of the general weighted model, is a not an elementary distinct model). Thus a new light is shed upon the Boolean $IR$ model, as follows: the classical Boolean $IR$ (see, e.g., Definition 4.36 or part 3.1) is a special case of the Weighted Boolean $IR$, and its underlying mathematical structure depends on an other particular model used for the document selecting condition. The weighted (i.e., non–Boolean) Weighted Boolean $IR$ is itself a distinct model of $IR$.

## 6.3    Bibliographical Remarks

For the traditional Boolean model of $IR$, see e.g. [van Rijsbergen, 1979; Salton and McGill, 1983; Kraft, 1985; Turtle and Croft, 1992; Bordogna and Pasi, 1993].

The Weighted Boolean $IR$ [see, e.g., Buell, 1982; Kraft and Buell, 1983; Salton, 1983; Kraft, 1985; Yager, 1987; Bordogna and Pasi, 1993; Bordogna, Carrara and Pasi, 1995; Kraft, Bordogna and Pasi, 1998] is an extension of the traditional model (based on Fuzzy Set Theory).

## 7.   NETWORK STRUCTURE OF INFORMATION RETRIEVAL MODELS

Figure 3.10 was a schematic of the traditional formal structure of *IR* models. In light of the mathematical foundation and theory given Chapter 4, the new formal structure of *IR* models becomes as shown in Figure 4.1.

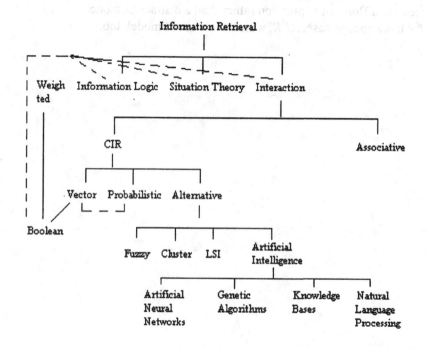

*Figure 4.1.* New structure of IR.

This diagram shows major changes as compared to the diagram of Figure 3.10.

Formally there are four models of *IR*: Information Logic *IR* (*ILIR*), Situation Theory *IR* (*STIR*), Interaction *IR* (*I²R*), and Weighted Booelan *IR* (*WBIR*).

It is possible to formally derive all models but the classical Boolean from *I²R*. Thus from a hierachical and formal point of view the structure of *IR* models has simplified considerably.

The Weighted Boolean $IR$ model is a distinct one, provided it is weighted (see text above). It has a special case, the classical Boolean model, when all the weights are exclusively Boolean. The classical Boolean model exhibits a duality in that: (1) it is a classical Boolean vector model owing to its document selection condition; but (2) in principle, for its document selection condition, any other model may be used. Because of this duality, the classical Boolean model is both 'less' and 'more' than any other model. In other words, it may be said that the classical Boolean $IR$ is more a method of processing a Boolean expression rather than a distinct $IR$ model.

$I^2R$ has a special case, $AI^2R$, which is a distinct model, too.

# Chapter 5

# RELEVANCE EFFECTIVENESS IN INFORMATION RETRIEVAL

## 1. RELEVANCE

It is commonly agreed that there exists a concept around — or based on — which the whole theory and practice of *IR* should (or could) be built — this notion is called *relevance*.

In general, the meaning of the word *relevance* is described as follows:

> "a state or quality of being to the purpose; a state or quality of being related to the subject or matter at hand" [The Cambridge English Dictionary, Grandreams Limited, London, English Edition, 1990]

Relevance is a complex and widely studied notion in several fields such as philosophy, psychology, communication theory, artificial intelligence, library science, and so on. And yet, it is not completely understood nor is it mathematically defined in an acceptable manner.

It plays a major role not only in *IR*, but e.g. in information science, too. Moreover, information science emerged *on its own* and *not as a part* of some other discipline, because scientific communication has to deal not with *any* kind of information but with *relevant* information. The creators of mathematical information theory, Shannon and Weaver [1949], begin their land mark book by pointing out that relevance is a central problem in communication: Is an American news program relevant to a Russian who does not speak English?

It is commonly accepted that when trying to understand the meaning of the concept of relevance, or when trying to interpret various meanings of this word, other concepts should be involved such as, for example, information, channel, message, meaning, etc.. Because a comforting definition of the concept of information, for example, is lacking — and whether this is possible at all is an open question [e.g. Shannon and Weaver, 1949; Devlin, 1991; Barwise, 1993; Mizzaro, 1997] —, there is little hope to have a definition for the notion of relevance.

All the previous mathematical models of *IR* assume that there is a relevance relationship between a query and documents. This relationship seems to be the most critical variable — both from theoretical and practical point of view — these models are confronted with [see e.g. Meadow, 1988; van Rijsbergen, 1989; Frants, 1991].

Relevance is considered to be a central concept in *IR* and hence a chief (theoretical and practical) modelling parameter in that *IR* is concerned with retrieving information relevant to a user's information need.

There has been a vast experimental and theoretical research and there is a vast literature on different aspects of relevancy in *IR*.

Major theoretical work so far has been done in e.g. [Maron and Kuhns, 1960; Goffman, 1964, 1970; Goffman and Nevill 1967; Cooper, 1968, 1971; Rocchio, 1971; Salton, 1971a; Negoita, 1973; Kochen, 1974; Radecki, 1976, 1977; Cooper and Maron, 1978; Kraft and Bookstein, 1978; Dillon and Desper, 1980; Bookstein, 1977, 1979, 1983; Buell and Kraft, 1981; Smeaton, 1984; Radecki and Kraft, 1985; Salton, Fox and Vorhees, 1985; Chang and Chen, 1987; Koll and Srinivasan, 1990; Bruza and van der Weide, 1991, 1992; Gordon and Lenk, 1991; Efthimiadis, 1993; Lalmas and van Rijsbergen, 1992, 1993; van Rijsbergen and Lalmas, 1996; Bruza, 1993; van Rijsbergen, 1986a, 1986b, 1989; Nie, 1988, 1989; Tiamiyu and Ajiferuke, 1988; Janes, 1994; Meghini, Sebastiani, Straccia and Thanos, 1993; Kraft et al., 1994; Sebastiani, 1994; Barry, 1994; Crestani and van Rijsbergen, 1995a, 1995b; Nie, Brisebois and Lepage, 1995; Lalmas, 1998; Bordogna and Pasi, 1996; Mizzaro, 1996, 1997; Dominich, 1994, 1997b, 1999a; Pasi and Marques Pereira, 1999; Heine, 1999a, 1999b].

The picture of relevance which unfolds from these researches reveals three components of relevance, as follows:

1. *Subjective*. The subjective component means that relevance is judged by (depends on) the user.
2. *A priori*. The *a priori* component means that (a measure of) relevance is built into the *IR* system, e.g., in the form of a similarity or probability measure.
3. *Emergent*. The emergent component means that relevance emerges or develops (users' relevance assessment may change) such as, for example, in a relevance feedback process, in a World Wide Web (WWW)

searching session, in a retrieval based on latent semantic indexing (*LSI*), or on interaction (where even documents which do not contain query terms may also be returned and found to be relevant).

Thus relevance proves to be a compound and dynamic category being a dynamic system of (at least) three components, as follows (Table 5.1):

*Table 5.1.* Relevance and its components.

| CATEGO-RY | COMPONENT | COMPONENT | COMPONENT | COMPO-NENT |
|---|---|---|---|---|
| relevance | subjective | a priori | emergent | ? |

The different components of relevance usually appear in a mixed combination in different *IR* models. For example, the similarity (vector) *IR* model typically contains a combination of the *a priori* and subjective components, a relevance feedback process typically is a combination of all three components. Table 5.2 shows a distribution of relevance components in *IR* models.

*Table 5.2.* Distribution of relevance components in *IR* models. Naturally, the subjective component occurs in every model. The parentheses contain an example as to how the respective components appear.

| IR Model | Subjective | A priori | Emergent |
|---|---|---|---|
| Boolean | yes | yes (literal match) | ? |
| Vector | yes | yes (similarity) | ? |
| Probabilistic | yes | yes (conditional) | yes (relevance feedback) |
| Cluster | yes | yes (similarity to centroid) | yes (cluster members) |
| Information Logic | yes | ? | yes (inference) |
| Situation | yes | ? | yes (flow) |
| Fuzzy Set | yes | yes (membership) | ? |
| Neural Networks | yes | yes (weights) | yes (activation) |
| Genetic Algorithms | yes | yes (fitness) | ? |
| Latent Semantic | yes | yes (similarity) | yes (k–factorisation) |
| Interaction | yes | ? | yes (recalled memories) |

## 2.    EFFECTIVENESS MEASURES

The *effectiveness* of an *IR* system means how well or badly it performs. The complexity of this task, as well as the compoud nature of effectiveness, is well reflected in research in recent years [e.g., Mizzaro, 1997; Allan, 1996; Belkin et al., 1996; Belkin and Koenemann, 1996; JASIS, 1994; Su, 1992, 1994, 1998].

The effectiveness is expressed by *effectiveness measures* elaborated based on different categories such as:

- Relevance
- Efficiency
- Utility
- User satisfaction

Within each category there are different specific effectiveness measures:

- Relevance (e.g., precision, recall, fallout)
- Efficiency (e.g., cost of search, amount of search time)
- Utility (e.g., worth of search results in some currency)
- User satisfaction (e.g., user's satisfaction with precision or an intermediary's understanding of a request).

Because of the central role played by relevance it has been and is used to work out specific measures of effectiveness for *IR* systems (which have been widely accepted), i.e. measures to express how well (or badly) an *IR* system performs. The following are traditional measures:

- *Precision*. Precision means the proportion of relevant documents out of those returned.
- *Recall*. Recall means the proportion of returned documents out of the relevant ones.
- *Fallout*. Fallout means the proportion of returned documents out of those nonrelevant.

Obviously, these measures are neither unmistakable nor absolute. (We should not, however, forget the partly emergent and probably not quantitative component of relevance [see, e.g., Heine, 1999a, 1999b; Dominich, 1994]; to quote Heine, "The concept of relevance does not have *a prior* existence, but is rather created 'on the fly', at least in some cases.") For instance, the estimation of recall requires the *a priori* (i.e., before retrieval) knowledge of the total number of relevant documents in the entire collection (for a given query). However paradoxical this may sound, it is an experimental result that users are more concerned with high recall than precision [see, e.g., Su, 1994].

Attempts to balance these measures have been made, and various other complementary or alternative measures have been elaborated.

Cooper [1968] suggests expected search length, i.e., the number of nonrelevant documents before finding the relevant ones.

Van Rijsbergen [1979] proposes a weighted combination of recall and precision:

$$1 - ((\alpha \cdot Precision \cdot Recall)/(\beta \cdot Precision + Recall)) \tag{5.1}$$

In [Bollmann–Sdorra and Raghanvan, 1993] another measure called $R_{norm}$ is suggested:

$$R_{norm} = 0.5 \cdot (1 + R^+ - I^-) \tag{5.2}$$

where $R^+$ denotes the number of times a relevant document occurs before a non–relevant one in the retrieval order, and $\Gamma$ is the number of times a non–relevant document occurs after a non–relevant one in the retrieval order.

Another standard effectiveness measure is referred to as the *harmonic mean*, $2\rho\pi/(\rho+\pi)$; see below.

There are other alternative measures, based on Fuzzy Set Theory [e.g. Ragade and Zunde, 1974; Buell and Kraft, 1981; Miyamoto, 1990].

In what follows the three widely accepted and used measures — i.e. precision, recall and fallout — will be considered, and their basic mathematical properties are recalled.

Mathematically these three measures can be dealt with as follows.

Let

- $\Delta \neq 0$ denote the total number of relevant documents to a query $q$;
- $|\mathfrak{R}(q)| = \kappa \neq 0$ denote the number of retrieved documents in response to $q$;
- $\alpha$ denote the number of retrieved and relevant documents.

It is reasonable to assume that $|D| = M > \Delta$.

The usual relevance effectiveness measures are defined now as follows.

*DEFINITION 5.1*
*Recall* $\rho$ is defined as $\rho = \alpha / \Delta$.

*DEFINITION 5.2*
*Precision* $\pi$ is defined as $\pi = \alpha / \kappa$.

*DEFINITION 5.3*
*Fallout* $\varphi$ is defined as $\varphi = (\kappa - \alpha) / (M - \Delta)$.

One can easily see that the following well known relations hold.

**PROPOSITION 5.1 [BUCKLAND AND GEY 1994]**
The ratio of recall and precision varies linearly with $\kappa$.

**Proof.** $\alpha = \rho\Delta = \pi\kappa \Rightarrow \rho / \pi = \kappa / \Delta$. ◆

Using experiments, Buckland and Gey [1994] elaborate polynomials for precision and recall using postulates based on empirical considerations. Their result is

$$P = N_{rel} \frac{R}{x} \qquad (5.3)$$

where $P$ means precision, $R$ means recall, $N_{rel}$ is the total number of relevant documents and $x$ is the number of retrieved documents. Observe that this rewrites as

$$\frac{R}{P} = \frac{x}{N_{rel}} \qquad (5.4)$$

which coincides with that of Proposition 1.
    As another property, it can be easily shown that:

**PROPOSITION 5.2**
Recall, precision and fallout satisfy the following relationship ($\rho \neq 0$, $\pi \neq 1$):

$$\frac{\varphi\pi}{\rho(1 - \pi)} = \frac{\Delta}{M - \Delta}$$

**Proof.**

$$\varphi = \frac{\kappa - \alpha}{M - \Delta} = \frac{\kappa - \rho\Delta}{M - \Delta} = \frac{\rho\Delta/\pi - \rho\Delta}{M - \Delta} = \frac{\rho\Delta(1 - \pi)}{\pi(M - \Delta)} \quad ◆$$

## 3. RELEVANCE FEEDBACK

We are going to show an underlying abstract mathematical structure behind *relevance feedback* (and hence probabilistic *IR*, too, where relevance feedback is typically used) which is a tool to improving effectiveness.

In principle, a typical retrieval scenario looks as follows:

1. USER has an
2. INFORMATION NEED (wants to find out something) which is expressed in a
3. REQUEST which is formalised into a
4. QUERY (a form 'understood' by the *IR* system) based on which
5. RETRIEVAL is performed, after which depending on
6. USER'S RELEVANCE JUDGEMENT re–iterate from point 2 or stop.

Obviously, relevance and its assessment appear in point 6 (subjective component) and in point 5 as well (*a priori* and emergent components). But also, relevance features in point 3, for example, namely, does the request reflect the information need?

In all basic model types of *IR*, when performing retrieval, some definition or interpretation of relevance is adopted based on which retrieval (point 5 in the scenario above) is conceived (see also Table 5.2):

– In the similarity model of *IR* (or vector *IR*), the similarity measure is the expression of a relevance. It is assumed (or accepted) that relevance is a dichotomic two–valued category, i.e., it either exists (relevance) or does not exist (irrelevance or nonrelevance).
– In the probabilistic *IR* the conditional probabilities are an expression of relevance. As above, relevance is dichotomic, but with a measure of dichotomy.
– In the logic models of *IR* van Rijsbergen's principle is a (logic) measure of a relevance.
– In the fuzzy model of *IR* there are different degrees of relevance. Relevancy is a multivalued non–dichotomic category.
– In the interaction model of *IR* the result of the interaction (memories) is an expression of relevance. It is suggested that relevance is an emergent category, too.

The user's judgement of relevance, point 6 in the scenario above, is a subjective momentum and thus probably inadequate for a mathematical definition.

However, relevance exists and is modelled by feeding it back into the *IR* system and using this feedback from the user to perform another — hopefully better — retrieval, and so on. This is called *relevance feedback*.

The technique of relevance feedback — to return documents that are likely to be (or are hopefully) more relevant —, can be used in several ways.

Query reformulation/expansion [see, e.g., Rocchio, 1971; Salton, 1971a; Salton et al. 1985; Dillon and Desper, 1980; Efthimiadis, 1993]. In the vector *IR* the old query vector $\mathbf{x}_{old}$ is replaced by a new query vector $\mathbf{x}_{new}$ by 'strengthening' it with terms from relevant documents:

$$\mathbf{x}_{new} = \alpha \cdot \mathbf{x}_{old} + \beta \cdot \sum_{k \in Rel\_docs} \mathbf{o}_k - \gamma \cdot \sum_{k \in NonRel\_docs} \mathbf{o}_k \qquad (5.5)$$

where $\mathbf{o}_k$ denotes document vector.

In the Boolean model of *IR*, a new query is constructed by taking a disjunction of prevalent terms from retrieved relevant documents.

Another technique is based on Bayes' Theorem [see e.g. van Rijsbergen, 1979]: recomputation of conditional probabilities, in probabilistic *IR* (see below).

In, e.g., [Radecki, 1977; Smeaton, 1984; Radecki and Kraft, 1985; Wong et al., 1993; Crestani, 1993; Bordogna and Pasi, 1996; Pasi and Pereira, 1999], relevance feedback is implemented using *AI* techniques. An *ANN* is constructed using the retrieved relevant documents and fuzzy term associations. Documents are represented as fuzzy sets, neurons are most significant terms in the retrieved relevant documents, and weights represent similarity measure values. A knowledge base is defined to decide when a document will be relevant.

Genetic Algorithms (*GA*) have also been used to implement relevance feedback [e.g. Yang and Korfhage, 1992; Petry et al. 1997; Kraft et al., 1994]. Mutation, crossover and fitness function techniques are used to obtain a fitter query (and document) from the initial one.

# 4.    MATHEMATICAL STRUCTURE IN RELEVANCE FEEDBACK

The process of relevance feedback can be iterated and thus a sequence of sets of (more relevant) retrieved documents is obtained.

It has been shown [see, e.g., Salton et al., 1985] that, in the vector *IR*, after a few iterations and query revisions process retrieval effectiveness does not improve. More mathematically speaking, iterated relevance feedback 'converges' to a limit which behaves like a 'stable' (optimal) point.

Thus the following general question can be raised:

*Given repeatedly applied relevance feedback processes. Is the sequence of sets of retrieved documents a chaotic sequence or does it exhibit some order or structure?*

In other words, is there any abstract common mathematical structure underlying iterated relevance feedback, in general?

This problem is dealt with in the following.

Whatever technique is used to implement relevance feedback, the general schematic of the process of relevance feedback is as follows:

1. The user is presented with a list of retrieved documents;
2. The user is asked to classify the retrieved documents according to his/her judgment of relevance;
3. This judgement of relevance information is used to retrieve a new set of documents (which are likely to better satisfy the user's information need).

Based on Proposition 5.1, it can be shown that by using recall, precision, and fallout a surface can be constructed with the following properties:

- It looks similar for every *IR* system which feeds back the user's relevance judgement (only its specific position, not its shape, in space varies depending on the total number of documents and the total number of relevant documents given a query);
- Each point on this surface corresponds to a 3–tuple (precision, recall, fallout), and thus to one retrieval process.

Thus a sequence of repeatedly applied relevance feedbacks corresponds to a sequence of points on this surface. Assuming that the sequence of relevance feedbacks improves the effectiveness of the *IR* system, i.e., it improves precision, recall, and fallout, the question of whether the corresponding sequence of points tends towards a specific point (corresponding to an optimum) arises, and if so which is that point? In other words, the sequence of retrievals means a 'walk' on a surface towards an optimal point.

In mathematical terms: is there an optimal (minimal or maximal) point that this sequence of points tends to?

As perhaps intuitively expected, the answer to this question will be yes: it will be formally shown that a repeatedly applied sequence of relevance feedbacks can be conceived as a recursive process which can be theoretically modelled by an important mathematical structure (called recursively enumerable set or Diophantine set). A noteworthy property of such a structure is that it has a fixed point. This point can be thought of as corresponding to a retrieval situation which cannot be improved anymore, hence the existence of an optimal point is formally granted and it can be computed numerically. This computation will also be performed.

This surface will be called an *effectiveness surface* which thus offers a geometrical view of relevance effectiveness in *IR* [Dominich, 1999a, 1999b]. At the same time it also reflects a theoretical dynamics among precision, recall and fallout during repeated relevance feedbacks.

Specific cases of the concepts of recursion, fixed point, Diophantine set, and level surface will be used, which are explained first below.

## 4.1    Recursion

Intuitively, recursion means to define a process (function, procedure, language) with reference to itself. Formally, recursion is a process whereby a function $f$, called a *primitive recursive function*, with $n+1$ variables is defined as follows:

$$f(x_1, x_2, ..., x_n, 0) = \alpha(x_1, x_2, ..., x_n), \tag{5.6}$$

$$f(x_1, x_2, ..., x_n, y + 1) = \beta(x_1, x_2, ..., x_n, y, f(x_1, x_2, ..., x_n, y)) \tag{5.7}$$

where $\alpha$ is a function with $n$ variables and $\beta$ a function with $n+2$ variables. In words, the function $f$ is given an initial value first represented by function $\alpha$, and then every next value of $f$ is defined by function $\beta$ using the previous value of function $f$.

Let us consider the particular case when $n = 1$. Then

$$f(x, 0) = \alpha(x) \tag{5.8}$$

$$f(x, y + 1) = \beta(x, y, f(x, y)) \tag{5.9}$$

where $x = x_1$ and hence the index can be omitted.

An example of recursion is the usual arithmetical addition:

$$f(x, 0) = x \tag{5.10}$$

$$f(x, y+1) = f(x, y) + 1 \tag{5.11}$$

## 4.2    Fixed point

Given a recursion with a corresponding $f$ (as above). Then there exists a situation when the value of $f$ coincides with the value on which $f$ is computed; symbolically

$$f(x) = x \qquad (5.12)$$

This value $x$ is referred to as a *fixed point*.

There are several *fixed point theorems* in recursion theory (e.g., Rogers Fixed–Point, First Recursion, and Second Recursion Theorems).

It is very important that a fixed point exists, and this corresponds to a situation where a retrieved set of documents is the same as that retrieved in the next step (after relevance feedback). This may, in principle, be interpreted as a case when effectiveness cannot be enhanced anymore. This concept and interpretation of a fixed point will be used here.

## 4.3    Diophantine Set

A subset $A \subseteq B$ is called a *recursively enumerable* (r.e.) *set* or *Diophantine set* (Diophantine structure) relative to $B$ if there exists a procedure (process, algorithm, program, language) which, when presented with an input $b \in B$, outputs 'yes' if and only if $b \in A$. When $b \notin A$ then the procedure does not end (undefined).

For example, the set of C language programs which halt on a given input is a r.e. set (relative to the set of all C language programs).

It is known from Recursion Theory that:

**THEOREM 5.1**
A set whose elements are given (generated) by a primitive recursive function is a Diophantine set. ♦

## 4.4    Level surface

Given a function

$$f: \mathbf{R}^3 \to \mathbf{R} \qquad (5.13)$$

$$f(x, y, z) \in \mathbf{R}, (x, y, z) \in \mathbf{R}^3 \qquad (5.14)$$

and a constant $c \in \mathbf{R}$. The set of points $(x, y, z)$ in space for which

$$f(x, y, z) = c \qquad (5.15)$$

is called a *level surface*.

For example, let

$$f(x, y, z) = x^2 + y^2 + z^2 \qquad (5.16)$$

and

$$c = 4 \qquad (5.17)$$

Then the level surface

$$x^2 + y^2 + z^2 = 4 \qquad (5.18)$$

is the sphere having its centre in the origin and radius equal to 2. All level surfaces in this example are spheres but with different radii.

## 4.5    Mathematical Structure of Relevance Effectiveness

Because recall $\rho$, fallout $\varphi$ and precision $\pi$ can take on different values, e.g., in a series of relevance feedbacks for the same query $q$, they can be thought of as values of variables in general as follows:

- $x$ for fallout;
- $y$ for precision;
- $z$ for recall.

Thus the left hand side of the expression in Proposition 5.2 corresponds to a three–variable function

$$f(x, y, z) = \frac{xy}{z(1-y)} \qquad (5.19)$$

The right hand side of the expression in Proposition 5.2, $\Delta/(M - \Delta)$, is constant for a query $q$ under consideration.

Thus one can consider all those values of function $f$, i.e., points in the three–dimensional Euclidean space that are equal to $\Delta/(M - \Delta)$. This is equivalent to defining a surface as follows:

*DEFINITION 5.4* [DOMINICH 1999A]

The level surface

$$\Sigma = \{ (x, y, z) \in \mathbf{R}^3 : \frac{xy}{z(1-y)} = \frac{\Delta}{M - \Delta} \}$$

is called the *effectiveness surface* of *IR* (corresponding to a given *q*).
Figure 5.1 shows a plot of the effectiveness surface.

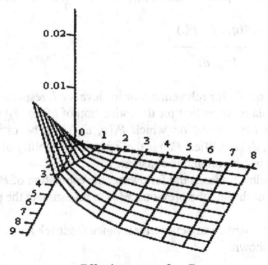

Effectiveness surface E.

E

*Figure 5.1.* The vertical axis corresponds to fallout, the axis to its right to precision, and the third axis corresponds to recall. It can nicely be seen that fallout increases when precision is low and recall is high. The lower on the surface, the higher precision and/or recall. The shape of the effectiveness surface is the same for every IR system, hence it is a typical surface. Its specific position in space varies with the total number of documents in the IR system and the total number of relevant documents given a query. The surface remains the same for the same query, and repeated relevance feedbacks for the same query means a walk on this surface towards its lowest point (this surface plays the role of a constraint in the optimisation of effectiveness).

Although the relevance feedback scenario is, in principle, the same in every IR model (as seen above), let us recall the relevance feedback process as used in the probabilistic model of IR. This does not mean any restriction upon the validity of arguments.

Let $q$ denote a query. In Probabilistic IR (PIR), the set $\mathcal{R}(q)$ of retrieved documents $d \in D$ in response to query $q$ contains those documents whose conditional probability of relevance $P(R|(q, d))$ exceeds that of non–relevance $P(I|(q, d))$ — Bayes' Decision Rule — and a real valued threshold (or cut off) $\tau \geq 0$, i.e.,

$$\mathcal{R}(q) = \{d \in D|\ P(R|(q, d)) \geq P(I|(q, d)), P(R|(q, d)) \geq \tau\} \qquad (5.20)$$

A basis to estimate the probabilities $P(.|(q, d))$ is offered by Bayes' Formula:

$$P(.|(q, d)) = \frac{P((q, d)|.)\ P(.)}{P(q, d)} \qquad (5.21)$$

where the dot (.) stands for relevance $R$ or irrelevance $I$, respectively.

Bayes' Formula requires that for the estimation of $P(.|(q, d))$ an initial set $\mathcal{R}_0(q)$ be known first, based on which $P((q, d)|.)$ can be estimated. The denominator $P(q, d)$ simplifies, $P(.)$ is the *a priori* probability of relevance $R$ and irrelevance $I$ in general and is constant.

Relevance feedback means the following: the estimation of $P(.|(q, d))$ can be iterated using each time the previous $\mathcal{R}(q)$ to re–estimate the probabilities $P((q, d)|.)$.

The following connection between relevance feedback and a Diophantine structure can be shown.

**THEOREM 5.2 [DOMINICH 1999A]**
Repeatedly applying relevance feedback yields a Diophantine set.

**Proof.** Given a query $q$. An initial set $\mathcal{R}_0(q)$ of retrieved documents is obtained first. Relevance feedback is repeatedly applied in consecutive steps $s = 1, 2, \ldots$.

At any step $s$, the set $\mathcal{R}_{s-1}(q)$ of the previous step is used to (re–)estimate the probabilities $P((q, d)|.)$ based on which the probabilities $P(.|(q, d))$ can be calculated — using Bayes' Formula — and a new set $\mathcal{R}_s(q)$ of retrieved documents is obtained.

Let $f(x, y)$ mean the newly retrieved set of documents $\mathcal{R}_s(q)$ at step $s$, where

a) $x$ is an integer variable corresponding to query $q$, and
b) $y$ is an integer variable symbolising step $s$ when probabilities $P(.|(q, d))$ are computed.

Let $\beta(x, y, f(x, y))$ represent the process of calculating, based on relevance feedback, the new probabilities $P(.|(q, d))$ and of retrieving a new set $\mathcal{R}_{s+1}(q)$ of documents, at step $s + 1$.

One can consider a sequence

$$\mathcal{R}_0(q), \; \mathcal{R}_1(q), \; \mathcal{R}_2(q), \; ..., \; \mathcal{R}_s(q), \; ... \tag{5.22}$$

of retrieved documents. Thus, one can define a function $f$ recursively as follows:

$$f(x, 0) = \alpha(x) \tag{5.23}$$

$$f(x, y + 1) = \beta(x, y, f(x, y)) \tag{5.24}$$

with the following meaning:

i)  at the initial step $s = 0$, i.e., $f(x, 0)$, an initial set $\mathcal{R}_0(q)$ is retrieved (using, e.g., a vector or interaction or another method), represented by $\alpha(x)$;
ii) then, at every next step $s + 1$, a new set $\mathcal{R}_{s+1}(q)$ is obtained, i.e., $f(x, y + 1)$, after repeatedly computing (based on relevance feedback using the previous $\mathcal{R}_s(q)$) the probabilities $P(.|(q, d))$ and performing a retrieval operation again, i.e. $\beta(x, y, f(x, y))$.

Because, formally, function $f$ is recursively defined (primitive recursive function), the series

$$\mathcal{R}_0(q), \; \mathcal{R}_1(q), \; \mathcal{R}_2(q), \; ..., \; \mathcal{R}_s(q), \; ... \tag{5.25}$$

forms a recursively enumerable (r.e.) set (relative to the power set $\wp(D)$ where $D$ denotes the set of documents to be searched), and as such it is a Diophantine set. ◆

As a consequence it is easy to show now that:

**THEOREM 5.3 [DOMINICH 1999A]**
Relevance feedback yields an optimal solution.

**Proof.**

The function $f$, being recursive, is computable, hence it has a fixed point (Rogers' Fixed Point Theorem, Philips, 1992). This means that there exists an index $s$ such that the same $\Re_s$ is obtained in a next step as that of the previous one, i.e., there is not any improvement, and thus this can be interpreted as an optimal situation. If one now assigns to the sets $\Re_s$ points on the effectiveness surface, a (optimal) fixed point acts as a limit towards this sequence of points tends to. ♦

## 5.     OPTIMALITY

It is easy to see that the following properties hold:

$$\rho = 0 \Leftrightarrow \pi = 0 \tag{5.26}$$

$$\pi = 1 \Leftrightarrow \varphi = 0 \tag{5.27}$$

$$\kappa = \Delta \Leftrightarrow (\rho = \pi = 1 \wedge \alpha = \kappa = \Delta \wedge \varphi = 0) \tag{5.28}$$

As is well known, an ideal (optimal) *IR* should be such that

$$\varphi = 0 \text{ and } \rho = 1 \tag{5.29}$$

Thus the following concept of optimality can be defined.

*DEFINITION 5.5*
An *IR* is said to be *optimal* if $\rho = 1$ and $\varphi = 0$.

An overall effectiveness (or performance) can be defined as a relative position, e.g., Euclidean distance or a cosine, to the ideal

$$I^* = (\varphi^*, \rho^*, \pi^*) = (0, 1, 1) \tag{5.30}$$

as follows:

*DEFINITION 5.6*
*Effectiveness (performance)* $\varepsilon$ of an *IR* is defined as the Euclidean distance between a current point $(\varphi, \rho, \pi)$ and the ideal $I^* = (0, 1, 1)$:

$$\varepsilon = \left(\varphi^2 + (1 - \pi)^2 + (1 - \rho)^2\right)^{1/2}$$

The smaller $\varepsilon$ the more effective *IR*. Thus, based on Theorem 5.3, repeated (iterated) relevance feedbacks can be conceived as an optimisation process which tries to minimise (i.e., enhance effectiveness) $\varepsilon$ taking into account an effectiveness surface. This can be numerically reformulated as the following non–linear minimisation problem with constraints:

Minimise

$$\left(\varphi^2 + (1 - \pi)^2 + (1 - \rho)^2\right)^{1/2} \tag{5.31}$$

subject to the following constraints:

$$\frac{\varphi\pi}{\rho(1 - \pi)} = \frac{\Delta}{M - \Delta}$$

$$0 < \rho \leq 1$$

$$0 \leq \pi < 1$$

$$0 \leq \varphi < 1$$

Alternatively, instead of the Euclidean distance, other measure for $\varepsilon$ may also be used, e.g. the cosine of the angle between the ideal $I^* = (0, 1, 1)$ and the current vector $\mathbf{o} = (\varphi, \pi, \rho)$:

$$\varepsilon = \frac{(I^*, \mathbf{o})}{\|I^*\| \, \|\mathbf{o}\|} = \frac{\pi + \rho}{(2(\varphi^2 + \pi^2 + \rho^2))^{1/2}} \tag{5.32}$$

In this case the higher $\varepsilon$ the more effective *IR* (the smaller the angle). The numerical counterpart is now as follows:

Maximise

$$\frac{\pi + \rho}{(2 \cdot (\varphi^2 + \pi^2 + \rho^2))^{1/2}} \tag{5.33}$$

subject to the following constraints:

$$\frac{\varphi\pi}{\rho(1-\pi)} = \frac{\Delta}{M-\Delta}$$

$$0 < \rho \leq 1$$

$$0 \leq \pi < 1$$

$$0 \leq \varphi < 1$$

Numerical solutions for both problems are as follows (using MathCAD 8.01 Plus Professional): $(\varphi, \pi, \rho) = (0.000001, 1, 0.9)$ for problem (5.31), and $(\varphi, \pi, \rho) = (0.0003, 0.8, 0.8)$ for problem (5.33).

Note

Taking into account 5.26—5.29, the above optimisation problems can take more simple forms, too.

Chapter 6

# FURTHER TOPICS IN INFORMATION RETRIEVAL

## 1. INFORMATION RETRIEVAL AND DECISION MAKING

In the recent past problems concerned with determining optimal solutions or assessing the suitability of products have led to the area of *multi–criteria decision making (MDM)*, which has two major branches:

* Multi–Objective Decision Making (MODM), which concentrates on continuous decision spaces and multiple objective functions;
* Multi–Attribute Decision Making (MADM), which focuses on discrete spaces and multiple criteria.

There is a vast literature on MDM. The interested reader is directed to e.g. [Hwang and Masud, 1979; Hwang and Yoon, 1981; Zimmermann, 1996; Miettinen, 1999].

A parallel between multi–criteria decision making and *IR* has been a research topic in recent years [see, e.g., Wong, Bollmann–Sdorra and Yao, 1991; Wong and Yao, 1991]. In principle, the parallel is based on the following: MADM is asking for the best alternative from a set of alternatives in the presence of multiple criteria. Making the following correspondances:

alternatives     documents

criteria     relevant, non–relevant, ...

the retrieval of a document in response to a query can be re–formulated in decision making terms as finding the best document (alternative) taking into account relevance (criterion).

In what follows, a formal comparison will be made between *MADM* and *IR* as a basis for a parallel and connection between the two areas, based on which the research mentioned can be formally and consistently introdcued.

An *MADM* problem can be formulated as follows [Zimmerman, 1996]:
Let

$$X = \{x_1, x_2, ..., x_j, ..., x_M\} \tag{6.1}$$

be a set of *alternatives*, and

$$G = \{\tilde{A}_1, \tilde{A}_2, ..., \tilde{A}_i, ..., \tilde{A}_C\} \tag{6.2}$$

a set of fuzzy sets representing *goals*. Each goal $\tilde{A}_i$ is assigned a *weight* $w_i$ representing its 'importance'. The attainment of goal $\tilde{A}_i$ by alternative $x_j$ is expressed by the degree of membership $\mu_i(x_j)$. A decision $\tilde{N}$ is defined as follows:

$$\tilde{N} = \underset{i=1, .C}{\otimes} (\tilde{A}_i)^{w_i} \tag{6.3}$$

where $\otimes$ denotes a confluence aggregator, e.g. fuzzy intersection or union. The optimal alternative $x^o$ is defined as follows:

$$\mu_N(x^o) = \underset{i=1, .C}{*} \mu_i(x_j) \tag{6.4}$$

where $*$ the corresponding fuzzy operator. (For an algorithmic solution to this problem see, e.g., Zimmermann [1996].)

The set $X$ of alternatives corresponds to set $D$ of objects, and the set $G$ of goals to set $A$ of criteria. The criteria do not have weights in *CIR*, or equivalently, all weights $w_i$ are set to 1 (in *MADM*). Given a *CIR*, i.e.:

$$D = \{\delta_j | \delta_j = \{t_k, \mu_j(t_k)\}\}, \quad \text{documents}, \tag{6.5}$$

$$A_q = \{\tilde{a}_i | \tilde{a}_i = ((q, \delta_j), \mu_i(q, \delta_j))\}, \quad \text{criteria}, \tag{6.6}$$

$$\mathcal{R}^i(q) = \{\delta | \mu_i(q, \delta) = \max_{1 \leq k \leq C} \mu_k(q, \delta)\} \cap a_{\alpha_i} \qquad (6.7)$$

A formal equivalence between *CIR* and *MADM* can be established as follows (The idea is to consider a table where the rows represent documents or alternatives, and the columns correspond to criteria or goals. Let us fix a criterion, say *i*, and assume that the weight is largest in the *i*th column, on every row. Then classical retrieval and decision are equivalent; they both mean the selection of the best alternative/document.):

**THEOREM 6.1**
If

a) the confluence aggregator $\otimes$ is the fuzzy union $\cup$,
b) $w_i = 1, \forall i$,
c) $D = X, A = G$, (*A* and *D* from *CIR*)
d) $\mu_i \geq \mu_k, \forall k$,
e) $a_{\alpha_i} = \max \{\{\mu_1, ..., \mu_i, ..., \mu_C\} \setminus \{\max \{\mu_1, ..., \mu_i, ..., \mu_C\}\}, \forall k$,

then *CIR* $\Leftrightarrow$ *MADM*.

**Proof.**
Based on points a) and b), one can write:

$$\otimes = \cup \Leftrightarrow \tilde{N} = \underset{i=1, ,C}{\otimes} (\tilde{A}_i)^{W_i} = \underset{i=1, ,C}{\cup} (\tilde{A}_i)^1 = \underset{i=1, ,C}{\cup} \tilde{A}_i$$

Hence, based on points d) and e), the following holds (*q* may be neglected as it is fixed, and $x = \delta$; see above):

$$\mathcal{R}^i = \{\delta \mid \max_{1 \leq k \leq C} \mu_k(\delta)\} \cap a_{\alpha_i}$$

$$= \{x_j \mid \mu_i(x_j)\} \cap a_{\alpha_i} = \{x_j \mid \mu_i(x_j)\}$$

$$= \underset{i=1, ,C}{\cup} \tilde{A}_i = \underset{i=1, ,C}{\otimes} (\tilde{A}_i)^{W_i} = \underset{i=1, ,C}{*} \mu_i(x_j) = \mu_N(x^o) \; \blacklozenge$$

## 2.       DATA FUSION

*Data fusion* means combining collections of documents provided by multiple *IR* systems. The importance of data fusion has been greatly improved by the advent of the World Wide Web (*WWW*) search engines on the Internet.

The idea of using multiple querying to enhance the effectiveness in *IR* has been in the focus of recent research [e.g., Belkin, Kantor, Cool and Quatrain, 1993; Voorhees, Gupta and Johnson–Laird, 1994, 1995; Lee, 1995; Croft, 1995; Kantor et al., 1995; Smeaton and Crimmins, 1999; Yager and Rybalov, 1998].

The data fusion problem can be formulated as follows.

Given a query $Q$. This is answered by several different *IR* systems using different *IR* models. Thus different collections of documents are obtained. The final answer to $Q$, i.e., the final set of documents, is obtained using these different collections of documents. Thus, data fusion is the problem of producing one (and possibly relevant) collection of documents (as a final answer to a query) by using or combining the different collections.

Formally, the concept of data fusion is defined using the notion of Information Retrieval Frame (*IRF*) $\Phi$, as follows.

*DEFINITION 6.1*
Given *IRF*s $\Phi_i = \langle D_i \cup \{Q\}, \Re_i \rangle$, $i = 1, 2, ..., n$. *Data fusion* is an *IRF* $\Phi = \langle D \subseteq \bigcup_{i=1, n} \Re_i(Q), \Re \rangle$.

Obviously the nature, structure, properties, and practical implementation of data fusion depend on how $\Re$ is defined.

Let

$$L_i = d_{i1}, d_{i2}, ..., d_{i\,|\Re_i(Q)|} \tag{6.8}$$

be an ordered list of the elements of set $\Re_i(Q)$. The ordering can be done in several ways.

Let all the frames $\Phi_i$ be *CIR*s. Then the different values provided by the $\mu_i$ functions (i.e. similarities $\sigma_i$ or probabilities $P_i$) can be viewed as values or weights (of relevance) assigned to each element $d_{ij}$.

A final list

$$L = d_1, d_2, ..., d_m \tag{6.9}$$

is a fused list where

$$m = b_1 + b_2 + \ldots + b_i + \ldots + b_n \tag{6.10}$$

and where $b_i \le |\Re_i(Q)|$, denotes the number of elements coming from list $L_i$. It is now easy to see the following connection between fusion and *SIR*:

**THEOREM 6.2**
If

$$\Phi_i = \langle D_i \cup \{Q\}, \Re_i \rangle, \quad i = 1, 2, \ldots, n$$

are *CIR*s, and the list $L$ is obtained by merging all lists $L_i$, then data fusion

$$\Phi = \langle D \subseteq \bigcup_{i=1, ,n} \Re_i(Q), \Re \rangle$$

is an *SIR* with similarity $\sigma$ defined using the scores and threshold $\tau$ so defined as to obtained the first $m$ elements.

**Proof.** $L$ is obtained by sorting all the elements in all $L_i$. Because $\sigma$ is already given, an appropriate threshold value $\tau$ needs be defined. ◆

Obviouly, fusion $\Phi$ can, in principle, be any other *IRF*. However, special procedures have been developed to implement $\Re$.
    Let

$$b_i = p_i b_1, \quad i = 2, .., n \tag{6.11}$$

where $p_i$ is a given proportion of elements $b_i$ relative to $b_1$ which is to be determined.
    Because

$$\sum_{i=1}^{n} b_i = m \tag{6.12}$$

one gets

$$m = b_1 + p_2 b_1 + p_3 b_1 + \ldots + p_n b_n$$

$$= b_1(1 + \sum_{i=2}^{n} p_i) \tag{6.13}$$

From this $b_1$ is obtained as follows

$$b_1 = m(1 + \sum_{i=2}^{n} p_i)^{-1} \qquad\qquad (6.14)$$

Thus all $b_i$, $i = 1, 2, ..., n$, can be computed.

The following fusion method is based on [Yager and Rybalov, 1998]. After selecting the first $r$ elements in $L$, there are $n_i$ elements left in $Li$. Each $L_i$ is associated the following value $V_i$:

$$V_i = \frac{n_i}{\sum_{j=1}^{n} n_j} \qquad\qquad (6.15)$$

As the next element to add to $L$, the top element left with the highest value $V_i$ is selected.

In order to handle ties between $L_i$s, an *indexing* is used as follows:

a) every $L_i$ is assigned an alphabetical name $Name_i$,
b) a number, called index($L_i$), is associated to each $L_i$ such that:

$$b_i > b_j \Rightarrow \text{index}(L_i) < \text{index}(L_j),$$
$$(b_i = b_j) \wedge (Name_i < Name_j) \Rightarrow \text{index}(L_i) < \text{index}(L_j)$$

If there is a tie, the element with the smallest *index* is selected.

EXAMPLE 6.1
Let the lists $L_i$ be as follows:

$L_1$: 1 2 3 4 5 6

$L_2$: a b c d

$L_3$: ! ?

We have $b_1 = 6$, $b_2 = 4$ and $b_3 = 2$. Firstly, the first $b_1 - b_2 = 6 - 4 = 2$ elements from $L_1$ are taken out and put into $L$:

$L = 5\ 6$

Thus, we have

$L_1$: 1 2 3 4

$L_2$: a b c d

$L_3$: ! ?

The top elements from $L_1$ and $L_2$ are then taken out in round robin fashion (the top element from $L_1$ first, then the top element from $L_2$, and so on), until both $L_1$ and $L_2$ contain $b_3 = 2$ elements:

$L =$ c 3 d 4 5 6

We have

$L_1$: 1 2

$L_2$: a b

$L_3$: ! ?

Next, the top elements from all three lists are taken out in round–robin fashion until all lists have been emptied:

$L =$ ! a 1 ? b 2 c 3 d 4 5 6

## 3.    INTERACTION AND SITUATION THEORY MODELS

Situation *IR* (*SITIR*) is a logical model for *IR* [e.g., Huibers and Bruza, 1996; van Rijsbergen and Lalmas, 1996] based on Situation Theory [Devlin, 1991]. A formal comparison between *SITIR* and $I^2R$ can be made.

The *SITIR* model is defined as follows. Given a set

$$D = \{d_k|\ k = 1, 2, ..., M\} \tag{6.16}$$

of *documents*. Let

$$S = \{s_k|\ k = 1, 2, ..., M\} \tag{6.17}$$

be an associated set of *situations*, where the situation

$$s_k = \{i_{kp}|\ i_{kp} = \langle\langle R_{kp}, a_{k1}, ..., a_{kj}, ..., a_{kn_p}, I_{kp}\rangle\rangle,\ \ j = 1, 2, ..., n_p,\ p = 1, ..., p_k\} \tag{6.18}$$

$$k = 1, 2, ..., M \tag{6.19}$$

is a set of *infons*.

In *SITIR* a counterpart of the concept of retrieval is suggested using the so called *situation aboutness*, as follows:

$$s_k \longmapsto s_i \ \Leftrightarrow \ \exists\ i_r \in s_i\ [\ s_k \models\rangle\ i_r\ ] \tag{6.20}$$

An element $v_k$ of $I^2R$ *D–Net* is assigned the following situation $s_k$

$$s_k = \{i_{kp}|\ i_{kp} = \langle\langle R_{kp}, \mathbf{w}_{kp}, I_{kp}\rangle\rangle,\ \forall\ p \neq k\ \} \tag{6.21}$$

Situation aboutness:

$$s_k \longmapsto s_i \ \Leftrightarrow \ \exists\ i_r \in s_i\ [\ s_k \models\rangle\ i_r\ ] \tag{6.22}$$

can then be defined as follows:

$$s_k \longmapsto s_i \ \Leftrightarrow \ \mathbf{w}_{jk} \neq \mathbf{0} \tag{6.23}$$

# APPENDIX 1

## Binary Similarity Information Retrieval (BSIR)

**I.** Define a set of objects $D = \{d_j | j=1,...,M\}$. Objects have (randomly generated) variable lengths and consist of (randomly generated) positive integers.

Number of objects $M$:      $M := 10$            $j := 0..\, M-1$

Variable length $L_j$ of object $d_j$:        $L_j := \text{floor}(\text{rnd}(8)) + 1$

$$L^T =$$

| | 0 | 1 | 2 | 3 | 4 | 5 | 6 | 7 | 8 | 9 | obj. number |
|---|---|---|---|---|---|---|---|---|---|---|---|
| 0 | 1 | 2 | 5 | 3 | 7 | 2 | 6 | 3 | 1 | 2 | obj. length |

Generate objects as consisting of (randomly generated) numbers.

$$d := \begin{vmatrix} \text{for } j \in 0..\, M-1 \\ \quad \text{for } i \in 0..\, L_j - 1 \\ \qquad d_{j,i} \leftarrow \text{floor}(\text{rnd}(10)+1) \\ d \end{vmatrix}$$

$d =$

| | 0 | 1 | 2 | 3 | 4 | 5 | 6 |
|---|---|---|---|---|---|---|---|
| 0 | 10 | 0 | 0 | 0 | 0 | 0 | 0 |
| 1 | 2 | 1 | 0 | 0 | 0 | 0 | 0 |
| 2 | 6 | 7 | 2 | 5 | 1 | 0 | 0 |
| 3 | 8 | 6 | 9 | 0 | 0 | 0 | 0 |
| 4 | 10 | 6 | 5 | 9 | 8 | 10 | 7 |
| 5 | 3 | 9 | 0 | 0 | 0 | 0 | 0 |
| 6 | 4 | 7 | 1 | 3 | 6 | 9 | 0 |
| 7 | 5 | 8 | 5 | 0 | 0 | 0 | 0 |
| 8 | 8 | 0 | 0 | 0 | 0 | 0 | 0 |
| 9 | 6 | 8 | 0 | 0 | 0 | 0 | 0 |

Each row is an object. The 0s do not belong to objects, they appear because of MathCAD's print. Objects are of variable lengths.

**II.** Define a set of identifiers $T = \{t_i | i = 1,...,N\}$. Identifiers are positive integers.

Number of identifiers $N$:      $N := 10$        $i := 0..\, N-1$

Generate identifiers:           $t_i := i + 1$

$$t^T =$$

| | 0 | 1 | 2 | 3 | 4 | 5 | 6 | 7 | 8 | 9 | id. number |
|---|---|---|---|---|---|---|---|---|---|---|---|
| 0 | 1 | 2 | 3 | 4 | 5 | 6 | 7 | 8 | 9 | 10 | identifier |

**III.** Assign binary vectors $X$:

$$x := \begin{array}{|l} \text{for } j \in 0..M-1 \\ \quad \text{for } k \in 0..N-1 \\ \quad\quad x_{j,k} \leftarrow 0 \\ \text{for } j \in 0..M-1 \\ \quad \text{for } k \in 0..N-1 \\ \quad\quad \text{for } i \in 0..L_j-1 \\ \quad\quad\quad x_{j,k} \leftarrow 1 \text{ if } t_k = d_{j,i} \\ x \end{array}$$

$x =$

|   | 0 | 1 | 2 | 3 | 4 | 5 | 6 | 7 | 8 | 9 |
|---|---|---|---|---|---|---|---|---|---|---|
| 0 | 0 | 0 | 0 | 0 | 0 | 0 | 0 | 0 | 0 | 1 |
| 1 | 1 | 1 | 0 | 0 | 0 | 0 | 0 | 0 | 0 | 0 |
| 2 | 1 | 1 | 0 | 0 | 1 | 1 | 1 | 0 | 0 | 0 |
| 3 | 0 | 0 | 0 | 0 | 0 | 1 | 0 | 1 | 1 | 0 |
| 4 | 0 | 0 | 0 | 0 | 1 | 1 | 1 | 1 | 1 | 1 |
| 5 | 0 | 0 | 1 | 0 | 0 | 0 | 0 | 0 | 1 | 0 |
| 6 | 1 | 0 | 1 | 1 | 0 | 1 | 1 | 0 | 1 | 0 |
| 7 | 0 | 0 | 0 | 0 | 1 | 0 | 0 | 1 | 0 | 0 |
| 8 | 0 | 0 | 0 | 0 | 0 | 0 | 0 | 1 | 0 | 0 |
| 9 | 0 | 0 | 0 | 0 | 0 | 1 | 0 | 1 | 0 | 0 |

Rows correspond to objects. Coloumns correspond to identifiers. 1 indicates the presence, 0 the absence of the corresponding identifier.

**IV.** Define similarity (Dice's coefficient) $m_1$

$$m1(u,v) := \begin{array}{|l} \text{num} \leftarrow 0 \\ \text{for } k \in 0..N-1 \\ \quad \text{num} \leftarrow \text{num} + u_k \cdot v_k \\ \text{den} \leftarrow 0 \\ \text{for } k \in 0..N-1 \\ \quad \text{den} \leftarrow \text{den} + u_k + v_k \\ \dfrac{2 \cdot \text{num}}{\text{den}} \end{array}$$

**V.** Define an object corresponding to a query $q$. This can be done in two ways: a) by fixing one of the objects in $D$ (which is not a real case) or b) by generating a new object (which is what happens in reality). From the point of view of mathematical structures, the two cases are eqauivalent to each other. Case b) will be considered.

Length of object query:                     $S := 3$          $s := 0 .. S - 1$

Generate object query:     $q_s := floor(rnd(10)) + 1$          $q^T = (6 \quad 2 \quad 5)$

Assign binary vector $xq$ to object-query q (as if an object-document):

$$xq := \begin{bmatrix} \text{for } k \in 0 .. N - 1 \\ \quad xq_k \leftarrow 0 \\ \text{for } k \in 0 .. N - 1 \\ \quad \text{for } i \in 0 .. S - 1 \\ \qquad xq_k \leftarrow 1 \text{ if } t_k = q_i \\ xq \end{bmatrix}$$

$xq^T =$

| | 0 | 1 | 2 | 3 | 4 | 5 | 6 | 7 | 8 | 9 |
|---|---|---|---|---|---|---|---|---|---|---|
| 0 | 0 | 1 | 0 | 0 | 1 | 1 | 0 | 0 | 0 | 0 |

The binary frame is formed by $X$ and $xq$ together:

**VI.** Define retrieval $R$:

$$\textbf{Retrieve1}(x, xq, \tau) := \begin{vmatrix} \text{for } j \in 0 .. M - 1 \\ \quad retrieved_j \leftarrow 1 \text{ if } m1\left(x^{<j>}, xq\right) > \tau \\ retrieved \end{vmatrix}$$

Obtain $R(q) = \{d \mid m1(q,d) > \tau\}$:          $\tau := 0.35$          $Rq := \textbf{Retrieve1}(x, xq, \tau)$

$Rq^T =$

| | 0 | 1 | 2 | 3 | 4 | 5 | 6 | 7 | 8 | 9 |
|---|---|---|---|---|---|---|---|---|---|---|
| 0 | 0 | 1 | 1 | 0 | 0 | 0 | 0 | 0 | 1 | 1 |

1 means object document retrieved,
0 means object document not retrieved.

**VII**. Using similarities m2 (Jaccard's coefficient) and m3 (Cosine measure):

$m2(u,v) :=$
$$\begin{array}{l} \text{num} \leftarrow 0 \\ \text{for } k \in 0..N-1 \\ \quad \text{num} \leftarrow \text{num} + u_k \cdot v_k \\ \text{den} \leftarrow 0 \\ \text{for } k \in 0..N-1 \\ \quad \text{den} \leftarrow \text{den} + \dfrac{(u_k + v_k)}{2^{u_k \cdot v_k}} \\ \dfrac{\text{num}}{\text{den}} \end{array}$$

$m3(u,v) :=$
$$\begin{array}{l} \text{num} \leftarrow 0 \\ \text{for } k \in 0..N-1 \\ \quad \text{num} \leftarrow \text{num} + u_k \cdot v_k \\ s1 \leftarrow 0 \\ s2 \leftarrow 0 \\ \text{for } k \in 0..N-1 \\ \quad s1 \leftarrow s1 + u_k \\ \quad s2 \leftarrow s2 + v_k \\ \dfrac{\text{num}}{\sqrt{s1} \cdot \sqrt{s2}} \end{array}$$

$\text{Retrieve2}(x, xq, \tau) :=$
$$\begin{array}{l} \text{for } j \in 0..M-1 \\ \quad \text{retrieved}_j \leftarrow 1 \ \ \text{if } m2\big(x^{<j>}, xq\big) > \tau \\ \text{retrieved} \end{array}$$

$\text{Retrieve3}(x, xq, \tau) :=$
$$\begin{array}{l} \text{for } j \in 0..M-1 \\ \quad \text{retrieved}_j \leftarrow 1 \ \ \text{if } m3\big(x^{<j>}, xq\big) > \tau \\ \text{retrieved} \end{array}$$

$\text{Retrieve2}(x, xq, \tau)^T = (0 \ \ 0 \ \ 0 \ \ 0 \ \ 0 \ \ 0 \ \ 0 \ \ 0 \ \ 0 \ \ 1)$

$\text{Retrieve3}(x, xq, \tau)^T =$

| | 0 | 1 | 2 | 3 | 4 | 5 | 6 | 7 | 8 | 9 |
|---|---|---|---|---|---|---|---|---|---|---|
| 0 | 0 | 1 | 1 | 0 | 0 | 0 | 0 | 0 | 1 | 1 |

Comparison of similarity values:     $i := 0 .. M - 1$

| i = | $m2\left[(x)^{<i>}, xq\right] =$ | $m1\left[(x)^{<i>}, xq\right] =$ | $m3\left[(x)^{<i>}, xq\right] =$ |
|---|---|---|---|
| 0 | 0.2 | 0.333 | 0.333 |
| 1 | 0.25 | 0.4 | 0.408 |
| 2 | 0.25 | 0.4 | 0.408 |
| 3 | 0 | 0 | 0 |
| 4 | 0.2 | 0.333 | 0.333 |
| 5 | 0.143 | 0.25 | 0.258 |
| 6 | 0.2 | 0.333 | 0.333 |
| 7 | 0.143 | 0.25 | 0.258 |
| 8 | 0.4 | 0.571 | 0.577 |
| 9 | 0.25 | 0.4 | 0.408 |

## Non-Binary Similarity Information Retrieval (NBSIR)

I., II. Given D and T as in Example 1.

III. Assigment of non-binary (frequency) vectors:

$$nx := \begin{vmatrix} \text{dummy} \leftarrow 0 \\ \text{for } j \in 0..N-1 \\ \quad \text{for } k \in 0..N-1 \\ \quad\quad x_{j,k} \leftarrow 0 \\ \text{for } j \in 0..M-1 \\ \quad \text{for } k \in 0..N-1 \\ \quad\quad \begin{vmatrix} \text{occur} \leftarrow 0 \\ \text{for } i \in 0..L_j - 1 \\ \quad \text{occur} \leftarrow \text{occur} + 1 \quad \text{if } t_k = d_{j,i} \\ x_{j,k} \leftarrow \dfrac{\text{occur}}{L_j} \end{vmatrix} \\ x \end{vmatrix}$$

nx =

|   | 0 | 1 | 2 | 3 | 4 | 5 | 6 | 7 | 8 | 9 |
|---|---|---|---|---|---|---|---|---|---|---|
| 0 | 0 | 0 | 0 | 0 | 0 | 0 | 0 | 0 | 0 | 1 |
| 1 | 0.5 | 0.5 | 0 | 0 | 0 | 0 | 0 | 0 | 0 | 0 |
| 2 | 0.2 | 0.2 | 0 | 0 | 0.2 | 0.2 | 0.2 | 0 | 0 | 0 |
| 3 | 0 | 0 | 0 | 0 | 0 | 0.333 | 0 | 0.333 | 0.333 | 0 |
| 4 | 0 | 0 | 0 | 0 | 0.143 | 0.143 | 0.143 | 0.143 | 0.143 | 0.286 |
| 5 | 0 | 0 | 0.5 | 0 | 0 | 0 | 0 | 0 | 0.5 | 0 |
| 6 | 0.167 | 0 | 0.167 | 0.167 | 0 | 0.167 | 0.167 | 0 | 0.167 | 0 |
| 7 | 0 | 0 | 0 | 0 | 0.667 | 0 | 0 | 0.333 | 0 | 0 |
| 8 | 0 | 0 | 0 | 0 | 0 | 0 | 0 | 1 | 0 | 0 |
| 9 | 0 | 0 | 0 | 0 | 0 | 0.5 | 0 | 0.5 | 0 | 0 |

IV. Given q as in Example 1.

V. Assignment of non-binary (frequency) vector to object query:

$$nxq := \begin{array}{|l} \text{for } k \in 0.. N-1 \\ \quad nxq_k \leftarrow 0 \\ \text{for } k \in 0.. N-1 \\ \quad \begin{array}{|l} occur \leftarrow 0 \\ \text{for } i \in 0.. S-1 \\ \quad occur \leftarrow occur + 1 \quad \text{if } t_k = q_i \\ nxq_k \leftarrow \dfrac{occur}{S} \end{array} \\ nxq \end{array}$$

| $nxq^T =$ | | 0 | 1 | 2 | 3 | 4 | 5 | 6 | 7 | 8 | 9 |
|---|---|---|---|---|---|---|---|---|---|---|---|
| | 0 | 0 | 0.333 | 0 | 0 | 0.333 | 0.333 | 0 | 0 | 0 | 0 |

VI. Define similarity (cosine measure):

$$nm3(u, v) := \begin{array}{|l} \left(\begin{array}{|l} num \leftarrow 0 \\ \text{for } k \in 0.. N-1 \\ \quad num \leftarrow num + u_k \cdot v_k \end{array}\right) \\ s1 \leftarrow 0 \\ s2 \leftarrow 0 \\ \text{for } k \in 0.. N-1 \\ \quad \begin{array}{|l} s1 \leftarrow s1 + u_k \\ s2 \leftarrow s2 + v_k \end{array} \\ \dfrac{num}{\sqrt{s1} \cdot \sqrt{s2}} \end{array}$$

VII. Obtain R(q), i.e., retrieval:

$$\text{Retrieve}(x, xq, \tau) := \begin{vmatrix} \text{dummy} \leftarrow 0 \\ \text{for } j \in 0 .. M - 1 \\ \quad \text{retrieved}_j \leftarrow 1 \quad \text{if } nm3\left(nx^{<j>}, nxq\right) > \tau \\ \text{retrieved} \end{vmatrix}$$

$\tau := 0.2$          $\text{Retrieve}(nx, nxq, \tau)^T = (0 \ \ 0 \ \ 1 \ \ 0 \ \ 0 \ \ 0 \ \ 0 \ \ 0 \ \ 1)$

## Comparison of BSIR and NBSIR

$\tau := 0.2$

$\text{Retrieve}(nx, nxq, \tau)^T = (0 \ \ 0 \ \ 1 \ \ 0 \ \ 0 \ \ 0 \ \ 0 \ \ 0 \ \ 1)$

$\text{Retrieve3}(x, xq, \tau)^T =$

$i := 0 .. M - 1$          **non – binary**                    **binary**

| $i =$ | $nm3\left[(nx)^{<i>}, nxq\right] =$ | $m3\left[(x)^{<i>}, xq\right] =$ |
|---|---|---|
| 0 | 0.179 | 0.333 |
| 1 | 0.199 | 0.408 |
| 2 | 0.204 | 0.408 |
| 3 | 0 | 0 |
| 4 | 0.047 | 0.333 |
| 5 | 0.041 | 0.258 |
| 6 | 0.067 | 0.333 |
| 7 | 0.031 | 0.258 |
| 8 | 0.2 | 0.577 |
| 9 | 0.084 | 0.408 |

## APPENDIX 2

# PROBABILISTIC INFORMATION RETRIEVAL

**I.** Set of object-documents D:

Number of objects: $M := 10$ $\quad\quad j := 0 .. M - 1$

Variable length L(j) of object d(j): $\quad L_j := \text{floor}\left(\text{rnd}(8)\right) + 1$

$$L^T = \begin{array}{c|c|c|c|c|c|c|c|c|c|c|} & 0 & 1 & 2 & 3 & 4 & 5 & 6 & 7 & 8 & 9 \\ \hline 0 & 3 & 8 & 6 & 6 & 6 & 1 & 4 & 8 & 1 & 4 \end{array}$$

Generate objects as consisting of numbers (instead of words, for instance) 0s do not belong to objects, 0s appear because of MathCAD's print:

$$d := \begin{array}{|l} \text{for } j \in 0 .. M - 1 \\ \quad \text{for } i \in 0 .. L_j - 1 \\ \quad\quad d_{j,i} \leftarrow \text{floor}\left(\text{rnd}(10) + 1\right) \\ d \end{array}$$

$$d = \begin{array}{c|c|c|c|c|c|c|c|c|} & 0 & 1 & 2 & 3 & 4 & 5 & 6 & 7 \\ \hline 0 & 3 & 2 & 6 & 0 & 0 & 0 & 0 & 0 \\ \hline 1 & 8 & 3 & 8 & 6 & 9 & 6 & 9 & 3 \\ \hline 2 & 4 & 1 & 4 & 4 & 3 & 2 & 0 & 0 \\ \hline 3 & 9 & 3 & 6 & 5 & 9 & 3 & 0 & 0 \\ \hline 4 & 7 & 10 & 8 & 9 & 2 & 1 & 0 & 0 \\ \hline 5 & 8 & 0 & 0 & 0 & 0 & 0 & 0 & 0 \\ \hline 6 & 10 & 8 & 7 & 7 & 0 & 0 & 0 & 0 \\ \hline 7 & 6 & 3 & 7 & 3 & 5 & 4 & 5 & 5 \\ \hline 8 & 1 & 0 & 0 & 0 & 0 & 0 & 0 & 0 \\ \hline 9 & 6 & 5 & 4 & 1 & 0 & 0 & 0 & 0 \end{array}$$

document_0

$\cdot$
$\cdot$
$\cdot$

document_9

**II.** Set of terms T (from query q):

$N := 9$ $\quad\quad\quad\quad i := 0 .. N \quad\quad\quad\quad t_i := i$

$$t^T =$$

| | 0 | 1 | 2 | 3 | 4 | 5 | 6 | 7 | 8 | 9 |
|---|---|---|---|---|---|---|---|---|---|---|
| 0 | 0 | 1 | 2 | 3 | 4 | 5 | 6 | 7 | 8 | 9 |

**III.** Define object query q:

Length of query: $s := 3$ $\quad s := 0 .. S - 1$

$q_s := floor(rnd(10)) + 1$ $\qquad q^T = [6 \ 2 \ 5]$

**IV.** Assigment of non binary (frequency) vectors to documents:

$$nx := \begin{bmatrix} \begin{aligned} & \text{for } j \in 0 .. M-1 \\ & \quad \text{for } k \in 0 .. N \\ & \qquad x_{j,k} \leftarrow 0 \\ & \text{for } j \in 0 .. M-1 \\ & \quad \text{for } k \in 0 .. N \\ & \qquad occur \leftarrow 0 \\ & \qquad \text{for } i \in 0 .. L_j - 1 \\ & \qquad \quad occur \leftarrow occur + 1 \ \text{ if } t_k = d_{j,i} \\ & \qquad x_{j,k} \leftarrow occur \\ & x \end{aligned} \end{bmatrix}$$

The document vectors are (in matrix form):

| | 0 | 1 | 2 | 3 | 4 | 5 | 6 | 7 | 8 | 9 |
|---|---|---|---|---|---|---|---|---|---|---|
| 0 | 0 | 0 | 1 | 1 | 0 | 0 | 1 | 0 | 0 | 0 |
| 1 | 0 | 0 | 0 | 2 | 0 | 0 | 2 | 0 | 2 | 2 |
| 2 | 0 | 1 | 1 | 1 | 3 | 0 | 0 | 0 | 0 | 0 |
| 3 | 0 | 0 | 0 | 2 | 0 | 1 | 1 | 0 | 0 | 2 |
| 4 | 0 | 1 | 1 | 0 | 0 | 0 | 0 | 1 | 1 | 1 |
| 5 | 0 | 0 | 0 | 0 | 0 | 0 | 0 | 0 | 1 | 0 |
| 6 | 0 | 0 | 0 | 0 | 0 | 0 | 0 | 2 | 1 | 0 |
| 7 | 0 | 0 | 0 | 2 | 1 | 3 | 1 | 1 | 0 | 0 |
| 8 | 0 | 1 | 0 | 0 | 0 | 0 | 0 | 0 | 0 | 0 |
| 9 | 0 | 1 | 0 | 0 | 1 | 1 | 1 | 0 | 0 | 0 |

$nx =$ (label for matrix, at row 4)

Assign binary vector to object query q:

$$nxq := \begin{vmatrix} \text{for } k \in 0 .. N \\ \quad x_k \leftarrow 0 \\ \text{for } k \in 0 .. N \\ \quad \text{for } i \in 0 .. S - 1 \\ \qquad x_k \leftarrow 1 \text{ if } t_k = q_i \\ x \end{vmatrix}$$

$nxq^T =$

|   | 0 | 1 | 2 | 3 | 4 | 5 | 6 | 7 | 8 | 9 |
|---|---|---|---|---|---|---|---|---|---|---|
| 0 | 0 | 0 | 1 | 0 | 0 | 1 | 1 | 0 | 0 | 0 |

**V.** Define similarity (based on dot product) function:

$$sim(u, v) := \begin{vmatrix} \begin{pmatrix} \begin{vmatrix} num \leftarrow 0 \\ \text{for } k \in 0 .. N \\ \quad num \leftarrow num + u_k \cdot v_k \end{vmatrix} \end{pmatrix} \\ s1 \leftarrow 0 \\ s2 \leftarrow 0 \\ \text{for } k \in 0 .. N \\ \quad \begin{vmatrix} s1 \leftarrow s1 + u_k \\ s2 \leftarrow s2 + v_k \end{vmatrix} \\ num \end{vmatrix}$$

VI. Obtain an initial set of retrieved documents R(q) to start with:

$$Retrieve(x, xq, \tau) := \begin{vmatrix} \text{for } j \in 0 .. M - 1 \\ \quad retrieved_j \leftarrow j \text{ if } sim\left[ (x^T)^{<j>}, xq \right] > \tau \\ retrieved \end{vmatrix}$$

$$\tau := \frac{N}{9} \qquad\qquad Retr\_docs := Retrieve(nx, nxq, \tau)$$

Modify (decrease) tau if none retrieved.

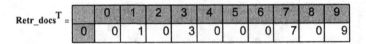

$$\text{Retr\_docs}^T =$$

| | 0 | 1 | 2 | 3 | 4 | 5 | 6 | 7 | 8 | 9 |
|---|---|---|---|---|---|---|---|---|---|---|
| 0 | 0 | 1 | 0 | 3 | 0 | 0 | 0 | 7 | 0 | 9 |

0 = not retrieved, otherwise retrieved

**VII.** Relevance feed back

Give the number of relevant documents:

$r := 1$          r = number of relevant documents minus 1

The relevant documents are:

$\text{Relevant}_0 := 1$          $\text{Relevant}_1 := 3$

Give the number of non-relevant documents:

$n := 1$          n = number of non-relevant documents minus 1

The irrelevant documents are (the rest of them):

$\text{Nonrelevant}_0 := 7$          $\text{Nonrelevant}_1 := 9$

**VIII.** Procedure to compute the probability that term $tk$ has $j$ occurrences in relevant/nonrelevant documents of Retr_Docs

$$
prob(RN, S) := \left| \begin{array}{l}
\text{for } k \in 0 \,.. \, N \\
\quad \left| \begin{array}{l}
\text{if } nxq_k \neq 0 \\
\quad \left| \begin{array}{l}
mr \leftarrow 0 \\
\text{for } i \in 0 \,.. \, r \\
\quad mr \leftarrow \left[ \left( nx^T \right)^{\left\langle \left( Relevant_i \right) \right\rangle} \right]_k \; \text{ if } mr < \left[ \left( nx^T \right)^{\left\langle \left( Relevant_i \right) \right\rangle} \right]_k \\
mnr \leftarrow 0 \\
\text{for } i \in 0 \,.. \, n \\
\quad mnr \leftarrow \left[ \left( nx^T \right)^{\left\langle \left( Nonrelevant_i \right) \right\rangle} \right]_k \; \text{ if } mnr < \left[ \left( nx^T \right)^{\left\langle \left( Nonrelevant_i \right) \right\rangle} \right]_k \\
m \leftarrow mr \\
m \leftarrow mnr \; \text{ if } mr < mnr \\
\text{for } j \in 0 \,.. \, m \\
\quad \left| \begin{array}{l}
u \leftarrow 0 \\
\text{for } i \in 0 \,.. \, S \\
\quad u \leftarrow u + 1 \; \text{ if } j = \left[ \left( nx^T \right)^{\left\langle \left( RN_i \right) \right\rangle} \right]_k \\
p_{k,j} \leftarrow \dfrac{u}{S+1}
\end{array} \right.
\end{array} \right.
\end{array} \right. \\
\quad p
$$

The probabilities for relevance are:

$pr := prob(Relevant, r)$

$$
pr = \begin{bmatrix}
0 & 0 & 0 & 0 \\
0 & 0 & 0 & 0 \\
1 & 0 & 0 & 0 \\
0 & 0 & 0 & 0 \\
0 & 0 & 0 & 0 \\
0.5 & 0.5 & 0 & 0 \\
0 & 0.5 & 0.5 & 0
\end{bmatrix}
$$

The probabilities for non-relevance are:

pnr := prob( Nonrelevant, n )

$$pnr = \begin{bmatrix} 0 & 0 & 0 & 0 \\ 0 & 0 & 0 & 0 \\ 1 & 0 & 0 & 0 \\ 0 & 0 & 0 & 0 \\ 0 & 0 & 0 & 0 \\ 0 & 0.5 & 0 & 0.5 \\ 0 & 1 & 0 & 0 \end{bmatrix}$$

**IX.** Computation of new document vectors (matrix) with probabilities:

$$nxp := \begin{vmatrix} nxp \leftarrow nx \\ C \leftarrow cols(pr) \\ \text{for } k \in 0 .. rows(pr) \\ \quad \text{for } j \in 0 .. M-1 \qquad\qquad \text{if } nxq_k \neq 0 \\ \quad\quad \text{if } nx_{j,k} \leq C \\ \quad\quad \begin{vmatrix} nxp_{j,k} \leftarrow \log\left(\dfrac{pr_{k,nx_{j,k}}}{pnr_{k,nx_{j,k}}}\right) & \text{if } \left(pr_{k,nx_{j,k}} \neq 0\right) \cdot \left(pnr_{k,nx_{j,k}} \neq 0\right) \\ nxp_{j,k} \leftarrow \infty & \text{if } \left(pr_{k,nx_{j,k}} \neq 0\right) \cdot \left(pnr_{k,nx_{j,k}} = 0\right) \\ nxp_{j,k} \leftarrow 0 & \text{if } \left(pr_{k,nx_{j,k}} = 0\right) \cdot \left(pnr_{k,nx_{j,k}} \neq 0\right) \end{vmatrix} \\ nxp \end{vmatrix}$$

The new matrix with probabiliy weights is:

$$
nxp = \begin{bmatrix}
0 & 0 & 1 & 1 & 0 & 1 \cdot 10^{307} & -0.3 & 0 & 0 & 0 \\
0 & 0 & 0 & 2 & 0 & 1 \cdot 10^{307} & 1 \cdot 10^{307} & 0 & 2 & 2 \\
0 & 1 & 1 & 1 & 3 & 1 \cdot 10^{307} & 0 & 0 & 0 & 0 \\
0 & 0 & 0 & 2 & 0 & 0 & -0.3 & 0 & 0 & 2 \\
0 & 1 & 1 & 0 & 0 & 1 \cdot 10^{307} & 0 & 1 & 1 & 1 \\
0 & 0 & 0 & 0 & 0 & 1 \cdot 10^{307} & 0 & 0 & 1 & 0 \\
0 & 0 & 0 & 0 & 0 & 1 \cdot 10^{307} & 0 & 2 & 1 & 0 \\
0 & 0 & 0 & 2 & 1 & 0 & -0.3 & 1 & 0 & 0 \\
0 & 1 & 0 & 0 & 0 & 1 \cdot 10^{307} & 0 & 0 & 0 & 0 \\
0 & 1 & 0 & 0 & 1 & 0 & -0.3 & 0 & 0 & 0
\end{bmatrix}
$$

**X.** Documents ranked in increasing order of their (likely) relevance:

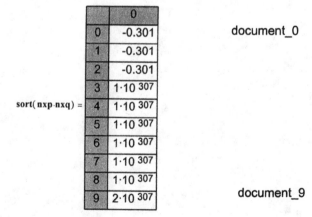

$$sort(nxp \cdot nxq) =$$

| | 0 |
|---|---|
| 0 | -0.301 |
| 1 | -0.301 |
| 2 | -0.301 |
| 3 | $1 \cdot 10^{307}$ |
| 4 | $1 \cdot 10^{307}$ |
| 5 | $1 \cdot 10^{307}$ |
| 6 | $1 \cdot 10^{307}$ |
| 7 | $1 \cdot 10^{307}$ |
| 8 | $1 \cdot 10^{307}$ |
| 9 | $2 \cdot 10^{307}$ |

document_0

document_9

Steps VII. onwards can be iterated. The sets Relevant and Nonrelevant may change at the user's wish.

# APPENDIX 3
## INTERACTION INFORMATION RETRIEVAL

## I. D-NET

**1.** Number of objects: $n := 5$      Maximum length of an object:    $l := 3$

**2.** Number of identifiers:   $vv := 5$      Generate objects (randomly) d:

$$i := 0 .. \, n - 1 \qquad L_i := \text{ceil}(\text{rnd}(l)) + 1$$

$$d_i := \left| \begin{array}{l} \text{for } j \in 0 .. \, L_i \\ \quad v_j \leftarrow \text{ceil}(\text{rnd}(vv)) + 1 \\ v \end{array} \right.$$

The objects are:

$$d_{0,0} = \begin{bmatrix} 2 \\ 5 \\ 3 \end{bmatrix} \quad d_{1,0} = \begin{bmatrix} 2 \\ 2 \\ 6 \end{bmatrix} \quad d_{2,0} = \begin{bmatrix} 2 \\ 2 \\ 4 \\ 5 \end{bmatrix} \quad d_{3,0} = \begin{bmatrix} 2 \\ 4 \\ 2 \\ 5 \end{bmatrix} \quad d_{4,0} = \begin{bmatrix} 4 \\ 6 \\ 6 \\ 4 \\ 4 \end{bmatrix}$$

**3.** Interconnect objects (weights change function L):

$$wijp := \left| \begin{array}{l} \text{for } i \in 0 .. \, n - 1 \\ \quad \left| \begin{array}{l} \text{for } j \in 0 .. \, n - 1 \\ \quad \left| \begin{array}{l} \text{if } j \neq i \\ \quad \left| \begin{array}{l} \text{for } p \in 0 .. \, L_j - 1 \\ \quad \left| \begin{array}{l} fijp \leftarrow 0 \\ v \leftarrow 0 \\ \text{for } k \in 0 .. \, L_i - 1 \\ \quad fijp \leftarrow fijp + 1 \ \text{if } (d_j)_p \stackrel{\blacksquare}{=} (d_i)_k \\ v_p \leftarrow \dfrac{fijp}{L_i} \end{array} \right. \\ vv_{i,j} \leftarrow v \end{array} \right. \end{array} \right. \end{array} \right. \\ vv \end{array} \right.$$

$$wikj := \left| \begin{array}{l} \text{for } i \in 0 .. \, n - 1 \\ \quad \left| \begin{array}{l} \text{for } j \in 0 .. \, n - 1 \\ \quad \left| \begin{array}{l} \text{if } j \neq i \\ \quad \left| \begin{array}{l} \text{for } k \in 0 .. \, L_i - 1 \\ \quad \left| \begin{array}{l} fikj \leftarrow 0 \\ v \leftarrow 0 \\ \text{for } p \in 0 .. \, L_j - 1 \\ \quad fikj \leftarrow fikj + 1 \ \text{if } (d_j)_p \stackrel{\blacksquare}{=} (d_i)_k \\ dfik \leftarrow 0 \\ \text{for } dd \in 0 .. \, n - 1 \\ \quad \text{for } p \in 0 .. \, L_{dd} - 1 \\ \quad\quad dfik \leftarrow dfik + 1 \ \text{if } (d_{dd})_p \stackrel{\blacksquare}{=} (d_i)_k \\ v_k \leftarrow fikj \cdot \log\left(\dfrac{2 \cdot n}{dfik}\right) \end{array} \right. \\ vv_{i,j} \leftarrow v \end{array} \right. \end{array} \right. \end{array} \right. \\ vv \end{array} \right.$$

**4.** Graphical representation of activity levels as sum of inputs. Compare this surface with that of D'-Net (next surface) to see the interaction. Merge links (for graphical representation purposes only):

$w := $ | for $ii \in 0.. n - 1$
     for $jj \in 0.. n - 1$
       $ww_{ii, jj} \leftarrow stack(wijp_{ii, jj}, wikj_{ii, jj})$ if $ii \neq jj$
   $ww$

$gd := $ | $k \leftarrow 0$
   for $i \in 0.. n - 1$
     for $j \in 0.. n - 1$
       if $i \neq j$
         $s \leftarrow 0$
         for $k \in 0.. \, length(w_{i, j}) - 1$
           $s \leftarrow s + (w_{i, j})_k$
         $v_{i, j} \leftarrow s$
   $v$

Compute $S_i$ :

$$gd = \begin{bmatrix} 0 & 0.5 & 0 & 0.5 & 0 \\ 0.155 & 0 & 0.31 & 1.31 & 0 \\ 0 & 0.667 & 0 & 1.065 & 1.129 \\ 0.155 & 0.976 & 0.643 & 0 & 0.333 \\ 0 & 0 & 0.898 & 0.398 & 0 \end{bmatrix}$$

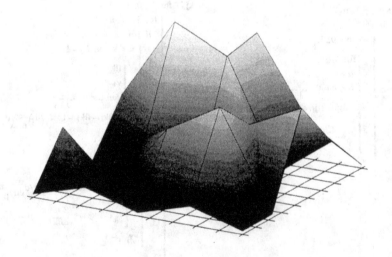

**gd**

## II. D'-NET

**1**. Generate query (randomly) as next object number n+1:

$L_n := ceil(rnd(1))$

(re)start here for
another retrieval

$d_n := \begin{vmatrix} \text{for } j \in 0..\ L_n - 1 \\ \quad v_j \leftarrow ceil(rnd(vv)) + 1 \\ v \end{vmatrix}$

$q := n$

$n := n + 1$

Object-query is:

$$d_q = \begin{bmatrix} 5 \\ 6 \\ 5 \end{bmatrix}$$

**2**. Incorporate query into D-Net (interaction, apply function L again):

$wwijp := \begin{vmatrix} \text{for } i \in 0..\ n-1 \\ \quad \begin{vmatrix} \text{for } j \in 0..\ n-1 \\ \quad \text{if } j \neq i \\ \quad \begin{vmatrix} \text{for } p \in 0..\ L_j - 1 \\ \quad \begin{vmatrix} fijp \leftarrow 0 \\ v \leftarrow 0 \\ \text{for } k \in 0..\ L_i - 1 \\ \quad fijp \leftarrow fijp + 1 \ \text{if } (d_j)_p = (d_i)_k \\ v_p \leftarrow \dfrac{fijp}{L_i} \end{vmatrix} \\ vv_{i,j} \leftarrow v \end{vmatrix} \\ vv \end{vmatrix}$

$wwikj := \begin{vmatrix} \text{for } i \in 0..\ n-1 \\ \quad \begin{vmatrix} \text{for } j \in 0..\ n-1 \\ \quad \text{if } j \neq i \\ \quad \begin{vmatrix} \text{for } k \in 0..\ L_i - 1 \\ \quad \begin{vmatrix} fikj \leftarrow 0 \\ v \leftarrow 0 \\ \text{for } p \in 0..\ L_j - 1 \\ \quad fikj \leftarrow fikj + 1 \ \text{if } (d_j)_p = (d_i)_k \\ dfik \leftarrow 0 \\ \text{for } dd \in 0..\ n-1 \\ \quad \text{for } p \in 0..\ L_{dd} - 1 \\ \quad dfik \leftarrow dfik + 1 \ \text{if } (d_{dd})_p = (d_i)_k \\ v_k \leftarrow fikj \cdot log\left(\dfrac{2 \cdot n}{dfik}\right) \end{vmatrix} \\ vv_{i,j} \leftarrow v \end{vmatrix} \\ vv \end{vmatrix}$

**4**. Graphical representation of activity levels as sum of inputs. Compare this surface with that of D-Net (previous surface) to see the interaction. Merge link (for graphical representation purposes only):

$w := \begin{vmatrix} \text{for } ii \in 0..\ n-1 \\ \quad \text{for } jj \in 0..\ n-1 \\ \quad ww_{ii,jj} \leftarrow stack\left(wwijp_{ii,jj}, wwikj_{ii,jj}\right) \ \text{if } ii \neq jj \\ ww \end{vmatrix}$

Compute $S_i$ :

```
gdp := | k ← 0
       | for i ∈ 0.. n − 1
       |   for j ∈ 0.. n − 1
       |     if i≠j
       |       | s ← 0
       |       | for k ∈ 0.. length(w_{i,j}) − 1
       |       |   s ← s + (w_{i,j})_k
       |       | v_{i,j} ← s
       | v
```

$$gdp = \begin{bmatrix} 0 & 0.5 & 0 & 0.5 & 0 & 1.704 \\ 0.234 & 0 & 0.468 & 1.468 & 0 & 0 \\ 0 & 0.667 & 0 & 1.144 & 1.288 & 0 \\ 0.234 & 1.135 & 0.801 & 0 & 0.333 & 0 \\ 0 & 0 & 0.977 & 0.477 & 0 & 0 \\ 1.269 & 0 & 0 & 0 & 0 & 0 \end{bmatrix}$$

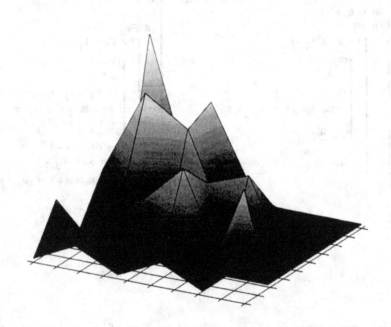

gdp

## III. Compute the size and intensity of interaction

$$
\text{par} := \left|
\begin{array}{l}
\text{size} \leftarrow 0 \\
\text{intensity} \leftarrow 0 \\
\text{for } i \in 0 \ldots n-2 \\
\quad \text{for } j \in 0 \ldots n-2 \\
\quad \text{if } i \neq j \\
\qquad \left|
\begin{array}{l}
\text{for } p \in 0 \ldots \text{if}\left(\text{length}\left(\text{wijp}_{i,j}\right) \leq \text{length}\left(\text{wwijp}_{i,j}\right), \text{length}\left(\text{wijp}_{i,j}\right) - 1, \text{length}\left(\text{wwijp}_{i,j}\right) - 1\right) \\
\quad \text{if } \left(\text{wijp}_{i,j}\right)_p \neq \left(\text{wwijp}_{i,j}\right)_p \\
\qquad \left|
\begin{array}{l}
\text{size} \leftarrow \text{size} + 1 \\
\text{intensity} \leftarrow \text{intensity} + \left(\text{wwijp}_{i,j}\right)_p - \left(\text{wijp}_{i,j}\right)_p
\end{array} \right. \\
\text{for } p \in 0 \ldots \text{if}\left(\text{length}\left(\text{wikj}_{i,j}\right) \leq \text{length}\left(\text{wwikj}_{i,j}\right), \text{length}\left(\text{wikj}_{i,j}\right) - 1, \text{length}\left(\text{wwikj}_{i,j}\right) - 1\right) \\
\quad \text{if } \left(\text{wikj}_{i,j}\right)_p \neq \left(\text{wwikj}_{i,j}\right)_p \\
\qquad \left|
\begin{array}{l}
\text{size} \leftarrow \text{size} + 1 \\
\text{intensity} \leftarrow \text{intensity} + \left(\text{wwikj}_{i,j}\right)_p - \left(\text{wikj}_{i,j}\right)_p
\end{array} \right.
\end{array} \right. \\
\left[ \begin{array}{c} \text{size} \\ \text{intensity} \end{array} \right]
\end{array}
\right.
$$

$\text{Size} := \text{par}_{0,0}$      $\text{Size} = 10$      $\text{Intensity} := \text{par}_{1,0}$      $\text{Intensity} = 1.188$

## IV. Retrieval (reverberative circle)

$$
\text{index}(v, e) := \left|
\begin{array}{l}
x \leftarrow 0 \\
\text{for } i \in 0 \ldots n-1 \\
\quad x \leftarrow i \text{ if } v_i = e \\
x
\end{array}
\right.
\qquad
\text{exist}(v, e, l) := \left|
\begin{array}{l}
x \leftarrow 0 \\
\text{for } i \in 0 \ldots l \\
\quad x \leftarrow i + 1 \text{ if } e = v_i \\
x
\end{array}
\right.
$$

$$\text{retrieval}(q) := \begin{vmatrix} qq \leftarrow 0 \\ {}^{o}qq \leftarrow q \\ \text{while } \text{exist}\left[o, \text{index}\left[\left(gdp^{T}\right)^{<{}^{o}qq>}, \max\left[\left(gdp^{T}\right)^{<{}^{o}qq>}\right]\right], qq\right] = 0 \\ \quad \begin{vmatrix} qq \leftarrow qq + 1 \\ {}^{o}qq \leftarrow \text{index}\left[\left(gdp^{T}\right)^{<\left({}^{o}qq - 1\right)>}, \max\left[\left(gdp^{T}\right)^{<\left({}^{o}qq - 1\right)>}\right]\right] \end{vmatrix} \\ j \leftarrow 0 \\ \text{for } i \in \text{exist}\left[o, \text{index}\left[\left(gdp^{T}\right)^{<{}^{o}qq>}, \max\left[\left(gdp^{T}\right)^{<{}^{o}qq>}\right]\right], qq\right] - 1 \,..\, qq \\ \quad \begin{vmatrix} v_{j} \leftarrow {}^{o}i \\ j \leftarrow j + 1 \end{vmatrix} \\ v \end{vmatrix}$$

$$\text{retrieval}(q) = \begin{bmatrix} 5 \\ 0 \end{bmatrix}$$

# REFERENCES

Agosti, M., Crestani, F. and Melucci, M. (1997). On the Use of Information Retrieval Techniques for the Automatic Construction of Hypertext. *Information Processing and Management*. 33(2): 133–144.

Aleksander, I. and Morton, H. (1990). *Neural Computing*. Chapman and Hall.

Alfred, G. (1993). *Modern differential geometry of curves and surfaces*. CRC Press, Inc.

Allan, J. (1996). Incremental Relevance Feedback. *In*: SIGIR '96 *Proceedings of the 19th International Conference on Research and Development in Information Retrieval*. Zurich, Switzerland, 270–278.

Allen, J. (1997). Building Hypertext Using Information Retrieval. *Information Processing and Management*. 33(2): 145–1159.

Arampatizis, A.T., Tsoris, T., Koster, C.H.A. and van der Weide, Th.P. (1998). Phase–Based Information Retrieval. *Information Processing and Management*. 34(6): 693–707.

Ausiello, G. (1975). *Complesita di calcolo delle funzioni*. Editore Boringhieri societa per azioni, Torino.

Baeza–Yates, R. and Ribeiro–Neto, B. (1999). *Modern Information Retrieval*. Addison Wesley Longman Publishing Co. Inc.

Bahydt, G. C. (1967). The effectiveness of non–user relevance assessments. *Journal of Documentation*. 23(2): 146–149.

Barry, C. L. (1994). User–defined relevance criteria: An exploratory study. *Journal of the American Society for Information Science*. 45(3): 149–159.

Barry, C.L. and Schamber, L. (1998). User's Criteria for Relevance Evaluation: A Cross-Situational Comparison. *Information Processing and Management*. 34(2/3): 219–236.

Barwise, J. (1993). Constraints, channels and the flow of information. *In*: Aczel, P., Asrael, D., Katagari. Y and Peters, S. (eds.) *Situation theory and its applications*. vol. 111, Stanford, CA: Stanford University, 3–27.

Beaulieu, M., Robertson, S.E. and Rasmussen, E. (1996). Evaluating interactive systems in TREC. *Journal of the American Society for Information Science*. 47(1): 85–94.

Belkin, N. J., Kantor, P., Cool, C. and Quatrain, R. (1993). Query combination and data fusion for information retreival. *In*: *Proceedings of the Second TREC Conference*, 25–44.

Belkin, N.J. and Koenemann, J. (1996). A case for interaction: A study of interactive information retrieval behavior and effectiveness. *In*: *Proceddings of the ACM SIG CHI Conference on Human Factors in Computing Systems*. New York, 205–212.

Belkin, N.J. et al. (1996). Using relevance feedback and ranking in interactive searching. *In*: Harman, D. (ed.) *TREC–4 Proceedings of Fourth Text Retrieval Conference*. Washington, D.C., 181–209.

Bernd, T. and Schmidt, S. (1988). Full Text Retrieval Based on Syntactic Similarities. *Information and System*. 13(1): 65–70.

Berry, M. W. and Dumais, S. T. (1994). Using Linear Algebra for Intelligent Information Retrieval. *Research Report*, Computer Science Department, University of Tennesse, Knoxville, USA.

Bodner, R. and Song, F. (1996). Knowledge–Based Approaches to Query Expansion in Information Retrieval. In: McCalla, G. (ed.), *Advances in Artificial Intelligence*, New York, Springer.

Bohr, N. (1928). Quantenpostulat und neue Entwicklungen der Atomistik. *Naturwissenschaften*. **16**: 245–263.

Bohr, N. (1958). *Atomphysik und mentschliche Erkenntniss*. Friedrich Vieweg & Sohn, Braunschweig.

Bollmann–Sdorra, P. and Raghavan, V.V. (1993). On the Delusiveness of Adopting a Common Space for Modelling Information Retrieval Objects: Are Queries Documents?.*Journal of the American Society for Information Science*. **44**(10): 579–587.

Bookstein, A. (1977). When the most "pertinent" document should not be retrieved – An analysis of the Swets model. *Information Processing and Management*. **13**(6): 377–383

Bookstein, A. (1979). Relevance. *Journal of the American Society for Information Science*. **30**: 269–273.

Bookstein, A. (1980). Fuzzy requests: an approach to weighted boolean searches. *Journal of the American Society for Information Science*, **31**(4): 240–247.

Bookstein, A. (1983). Information retrieval: A sequential learning process. *Journal of the American Society for Information Science*. **34**: 331–342.

Bordogna, G. and Pasi, G. (1993). A Fuzzy Linguistic Approach Generalizing Boolean Information Retrieval: A Model and Its Evaluation. *Journal of the American Society for Information Science*. **44**(2): 70–82.

Bordogna, G. and Pasi, G. (1995). Linguistic aggregation operators in fuzzy information retrieval. *International Journal of Intelligent Systems*, **10**(2): 233–248.

Bordogna, G. and Pasi, G. (1996). A user adaptive neural network supporting a rule–based relevance feedback. *Fuzzy Sets and Systems*. **82**: 201–211.

Bordogna, G., Carrara, P. and Pasi, G. (1991). Query term weights as constraints in fuzzy information retrieval. *Information Processing and Management*, **27**(1): 15–26.

Bozkaya, T. and Ozsoyoglu, M. (1997). distance–based inexing for high–dimensional metric spaces. In: *Proceedings of the 1997 ACM SIGNOID Interantional Conference on Management of Data*. Tucson, AZ, 357–368.

Brin, S. (1995). Near neighbor search in large metric spaces. In: *Proceedings of the 21st VLDB Interantional Conference*. Zürich, Switzerland, 574–584.

Brooks, H. M. (1987). Expert Systems and Intelligent Information Retrieval. *Information Processinn and Management*, **23**(4): 367–382.

Bruza, P. D. (1993). *Stratified information disclosure: A synthesis between hypermedia and information retrieval*. PhD Dissertation, University of Nijmegen, Nijmegen, The Netherlands.

Bruza, P. D. and van der Weide, T. P. (1991). The modelling and retrieval of documents using index expression. *SIGIR Forum*. **25**(2): 91–103.

Bruza, P. D. and van der Weide, T. P. (1992). Stratified hypermedia structures for information disclosure. *The Computer Journal*. **35**(3): 208–220.

Bruza, P.D. and van Linder, B. (1996). Preferential Models of Query by Navigation. *Research Report*. School of Information Systems and Research Data Network CRC, Queensland University of Technology, Australia, 1–19.

Buckland, M. and Gey, F. (1994). The Relationship Between Recall and Precision. *Journal of the American Society for Information Science*. **45**(1): 12–19.

Buckland, M.K. (1997). What is a document? *Journal of the American Society for Information Science*. **48**(9): 804–809.

Buell, D. A. (1982). An analysis of some fuzzy subset applications to information retrieval systems. *Fuzzy Sets and Systems*, **7**(1): 35–42.

Buell, D. A. (1985). A problem in information retrieval with fuzzy sets. *Journal of the American Society for Information Science*, **36**(6): 398–401.

Buell, D. A. and Kraft, D. H. (1981). Performance measurement in a fuzzy retrieval environment. *In*: *Proceedings of the Fourth International Conference on Information Storage and Retrieval. ACM/SIGIR Forum*, **16**(1): 56–62.

Burgin, R. (1999). The Monte Carlo Method and the Evaluation of Retrieval System Performance. *Journal of the American Society for Information Science*. **50**(2): 181–193.

Cahoon, B. and McKinley, K. (1996). An Architecture for Distributed Information Retrieval. *In*: *ACM SIGIR Proceedings of the 19th International Conference on Research and Development in Information Retrieval*. Zurich, Switzerland, 110–118.

Caid, W.R., Dimais, S.T. and Galiant, S.I. (1995). Learned Vector Space Models for Document Retrieval. *Information Processing and Management*. **31**(3): 419–429.

Callan, J.P., Croft, W.–B., Harding, S.–M. (1992). The INQUERY retrieval system. *In*: *Proceedings of the 3rd DEXA*. 78–83.

Carrick, C. and Watters, C. (1997). Automatic Association of News Items. *Information Processing and Management*. **33**(5): 615–632.

Cater, S. C. and Kraft, D. H. (1987). TIRS: A topological information retrieval system satisfying the requirements of the Waller–Kraft wish list. *In*: *Proceedings of the Tenth Annual International Conference Research and Development in Information Retrieval*. New Orleans, LA, 171–180.

Cater, S. C., Blane, K. and Harvel, L. (1989). Construction and evaluation of a prototype topological information retrieval system. *In*: *Proceedings of the 1989 IEEE Southeastcon*. Columbia, SC, 1336–1340.

Chang, S. C. and Chen, W. C. (1987). "And–less" retrieval toward perfect ranking. *In*: *Proceedings of the 50th Annual Meeting of the American Society for Information Science. Information: The Transformation of Society*. Boston, Medford, MA, NJ, 30–35.

Cheeseman, P., Kanefsky, B. and Taylor, W.M. (1991). Where the *Really* Hard Problems Are. *In*: Mylopoulos, J. and Reiter, R. (eds.) *Proceedings of the IJCAI–91*. Morgan Kaufmann, San Mateo, CA, 331–337.

Chellas, B. F. (1980). *Modal logic*. Cambridge University Press, Cambridge, MA.

Chen, J., Mikulcic, A. and Kraft, D.H. (1998). An integrated Approach to information Retrieval with Fuzzy Clustering and Fuzzy Inferencing. *In*: Pons, O., Amparo Vila, M. and Kacprzyk, J. (eds.) *Knowledge Management in Fuzzy Databases*. Physica Verlag, Heidelberg, Germany.

Chen, M, (1995). Machine Learning for Information Retrieval: Neural Networks, Symbolic Learning and Genetic Algorithms. *Journal of the American Society for Information Science*. **46**(3): 194–216.

Chen, P.S. (1994). On Inference Rules of Logic–Based Information Retrieval Systems. *Information Processing and Management*. **30**(1): 43–59.

Chiaramella, Y. (1992). About Retrieval Models and Logic. *The Computer Journal*. **35**(3): 233–241.

Chiaramella, Y. (1996). Our experience in logical *IR* modeling. *In*: Crestani, F. and Lalmas, M. (eds.) *Proceedings of Workshop on Logical and Uncertainty Models for Information Retrieval*. University of Glasgow, Glasgow, Scotland, 1–8.

Ciaccia, P., Patella, M. and Zezula, P. (1997). M–tree: An efficient access method for similarity search in metric spaces. *In*: *Proceedings of the 23rd VLDB International Conference*. Athens, Greece, 426–435.

Ciaccia, P., Patella, M. and Zezula, P. (1999). Processing Complex Similarity Queries with Distance–based Access Methods. *Research Report*. University of Bologna, Italy, CNR Pisa, Italy.

Cohen, W. W. and Fan, W. (1998). Learning page–independent heuristics for extracting data from Web pages. http://www.decweb.ethz.ch/WWW8

Cole, C. (1998). Intelligent Information Retrieval: Diagnosing Information Need. Part 1. The Theoretical Framework for Developing and Intelligent IR Tool. *Information Processing and Management.* 34(6): 709–720.

Cole, C., Cantero, P. and Suave, D. (1998). Intelligent Information Retrieval: Diagnosing Information Need. Part II.Uncertainty Expansion in a Prototype of a Diagnostic IR Tool. *Information Processing and Management.* 34(6): 721–737.

Cooper , W.S. and Maron, M.E. (1978). Foundation of probabilistic and utility–theoretic indexing . *Journal of the Association for Computing Machinery.* 25: 67–80.

Cooper, W. S. (1968). Expected search length: a single measure of retrieval effectiveness based on the weak ordering action of retrieval systems. *American Documentation.* 19: 30–41.

Cooper, W. S. (1971). A definition of relevance for information retrieval. *Information storage and retrieval*, 7(1): 19–37.

Cooper, W.S. (). Some inconsistencies and misidentified modelling assumptions in probabilistic information retrieval. *ACM Transactions On Information Systems.* 13(1): 100–111.

Copson, E.T. (1975). *Partial Differential Equations.* Cambridge University Press.

Crestani, F. (1993). Learning strategies for an adaptive information retrieval system using neural networks. In: *Proceedings of the IEEE Internatiuonal Conference on Neural Networks*, S. Francisco, CA., USA, March, 244–249.

Crestani, F. and van Rijsbergen, C.J. (1995a). Information retrieval by logical imaging. *Journal of Documentation.* 51: 3–17.

Crestani, F. and van Rijsbergen, C.J. (1995b). Probability kinematics in information retrieval. In: *ACM SIGIR 18th International Conference on Research and Development in Information Retrieval.* Seattle, WA, 291–299.

Crestani, F. and van Rijsbergen, C.J. (1996). Information retrieval by imaging. *In:* Leon, R. (ed.) *Information Retrieval New Systems and Current Research. Proceedings of the 16th Research Colloquim of the British Computer Society Information Retrieval Specialist Group.* Taylor Graham, Drymen, 47–67.

Cringean, J. K., England, R., Manson, G. A. and Willett, P. (1991). Network Design for the implementation of text searching using a multicomputer. *Information Processing and Management.* 27(4): 259–265.

Croft, W. B. and Lewis, D. D. (1987). An Approach to Natural Language Processing for Document Retrieval. *In: Proceedings of tenth Annual International ACM SIGIR Conference on research and Development in Information Retrieval.* New Orleans, LA, June 3–5, 26–32.

Croft, W. B. and Thompson, R. H. (1987). $I^3R$: A new approach to thedesign of document retrieval systems. *Journal of the American Society for Information Science.* 38: 389–404.

Croft, W. B. and Turtle, H. (1989). A Retrieval Model for Incorporating Hypertext Links. *In: Proceedings of the Second ACM Conference on Hypertext, Hypertext '89.* Pittsburg, USA, 213–224.

Croft, W.B. (1993). Knowledge–Based and Statistical Approaches to Text Retrieval. *IEEE Expert.* April, 8–12.

Croft, W.B. (1995). Effective Text Retrieval Based on Combining Evidence from the Corpus and User. *IEEE Expert.* December, 10(4): 59–63.

Croft, W.B. and Ponte, J. (1998). A Language Modelling Approach to Information Retrieval. *Proceddings of ACM SIGIR.* 275–281.

Croft, W.B., Harding, S.M. and Weir, C. (1997). Probabilistic Retrieval of OCR Degraded Text Using N–Grams. *In*: Peters, C. and Thanos, C. (eds.) *Research and Advances Technology for Digital Libraries*. 345–359.

Crouch, C. J. and Yang, B. (1992). Experiments in Automatic Statistical Thesaurus Construction. *Proceedings of the Fifteenth Annual International ACM SIGIR Conference on Research and Development in Information Retrieval*, 77–87.

Csaszar, A. (1978). *General Topology*. Akademiai Kiado, Disquivitiones Mathematicae Hungaricae, Budapest, Hungary.

da Silva, W.T. and Milidiu, R.L. (1993). Belief Function Model for Information Retrieval. *Journal of the American Society for Information Science*. 44(1): 10–18.

Daniels, J.J. and Rissland, E.L. (1995). A case–based approach to intelligent information retrieval. *In*: *ACM SIGIR Proceedings of the 18th International Conference on Research and Development in Information Retrieval*. 224–238.

Deerwester, S., Dumais, S., Furnas, G., Landauer, T., and Harshman, R. (1990). Indexing by latent semantic analysis. *Journal of the American Society for Information Science*, 41: 391–407.

Devaney, M. and Ram, A. (1996). Dynamically adjusting concepts to accomodate changing contexts. *In*: *Proceedings of the ICML–96 Workshop on Learning in Context–Sensitive Domains*. Bari, Italy, 6–14.

Devlin, K. (1991). *Logic and Information*. Cambridge University Press, Cambridge, Great Britain.

DeWilde, F. (1997). *Neural Network Models*. Springer Verlag.

Dey, D. and Sarkar, S. (1996). A probabilistic relational model and algebra. *ACM TOIS*. 21(3): 339–369.

Dillon, M. and Desper, J. (1980). The use of automatic Relevance Feedback in Boolean retrieval Systems. *Journal of Documentation*. 36(3): 197–208.

Dillon, M. and Gray, A. S. (1983). FASIT: A fully automatic syntactically based indexing system. *Journal of the American Society for Information Science*. 34: 99–108.

Dominich, S. (1990a). *User Modelling and Database Selection In Information Retrieval*. Technical Research Report 280, Institute for Information Processing, Joanneum Research Graz, Technical Univeristy Graz, Austria.

Dominich, S (1990b). Database Selection and Stereotypes. In: Haase, V. and Zinterhof, L. (eds.) *Proceedings of the FIT '90 Conference: Future Trends in Information Technology*, Oldenbourg, Wien, 26–28 September, Salzburg,Austria, 33–42.

Dominich, S. (1992). Die 'Kopenhagener Interpretation' zur Behandlung von Relevanz und Bedeutung im Information Retrieval. (The 'Copenhagen Interpretation' to Handle Relevanvy and Meaning in Information Retrieval.) *Symposium on Informatics*. Technical University Clausthal–Zellerfeld, Institute for Informatics, Germany, May 29 (lecture in German).

Dominich, S. (1993a). Artificial Intelligence In The Mathematical Models Of Information Retrieval. *In*: Koch, P. (ed.) *Proceedings of the Third Conference On Artificial Intellignce*. Budapest, John von Neumann Society for Computer Science, April 6–8, 221–230.

Dominich, S. (1993b). *The Formulation of the Interaction Information Retrieval Model As A New And Complementary Framework For Information Retrieval*. Ph.D. Thesis, Hungarian Academy of Sciences, Budapest, Hungary (in English).

Dominich, S. (1994). Interaction Information Retrieval. *Journal of Documentation*. 50(3): 197–212.

Dominich, S. (1997a). The Interaction–Based Information Retrieval Paradigm. *In*: Kent, A. and Williams, J. G. (eds.) *Encyclopedia of Computer Science and Technology*. Vol. 37, Suppl. 22, Marcel Dekker, Inc., New York Basel Hong Kong, 175–192.

Dominich, S. (1997b). The Interaction–Based Information Retrieval Paradigm. *In*: Kent, A. (ed.) *Encyclopedia of Library and Information Science*. Vol. 59, Suppl. 22, Marcel Dekker, Inc., New York Basel Hong Kong, 218–238.

Dominich, S. (1998). An $I^2R$ (Interaction Information Retrieval) Pre–processor for Relevance Feedback. *Technology Letters*. **2**(1): 5–18.

Dominich, S. (1999a). A geometrical view of relevance effectiveness in information retrieval. *In*: Crestani, F. and Lalmas, M. (eds.) *Proceedings of Workshop on Logical and Uncertainty Models for Information Systems*. University College, London, United Kingdom, 5th July, 12–22.

Dominich, S. (1999b). Mathematical Foundation of Information Retrieval. *Technology Letters*. **3**(1): 5–19.

Dominich, S. (1999c). Associative Database for Information Retrieval. *Proceedings of the 3rd Austrian–Israeli Technion Symposium cum Industrial Forum "Technology for Peace – Science and Mankind", Software for Communication Technology*. Austrian Technion Society, RISC Hagenberg Castle, Linz, Austria, 26–27 April, 205–209.

Dominich, S. (2000a). Foundation of Information Retrieval. *Mathematica Pannonica*, **11**(1): 137–153.

Dominich, S. (2000b). A Unified Mathematical Definition of Classical Information Retrieval. *Journal of the American Society for Information Science*, **51**(7): 614–625.

Dominich, S. (2000c). Formal Foundation of Information Retrieval. *Technology Letters*: ACM SIGIR 2000 Workshop on Mathematical/Formal Methdos in Information Retrieval, **4**(1): 8–16

Dubois, D., Prade, H., and Tstemale, C. (1988). Weighted fuzzy pattern matching. *Fuzzy Sets and Systems*, **28**:313–331.

Dunlop, M. D. and van Rijsbergen, C. J. (1993). Hypermedia and Free Text Retrieval. *Information Processing and Management*. **29**(3): 287–298.

Efthimiadis, E. N. (1993). A user–centered evaluation of ranking algorithms for interactive query expansion. *In*: *Proceedings of the sixteenth Annual International ACM/SIGIR Conference on Research and Development in Infromation Retrieval*. Pittsburgh, PA, 146–159.

Efthimiadis, E. N. (1995). User Choices: A New Yardstick for the Evaluation of Ranking Algorithms for Interactive Query Expansion. *Information Processing and Management*. **31**(4): 605–620.

Egghe, L. and Rousseau, R. (1988). A Theoretical Study of Recall and Precision Using a Topological Approach to Infromation Retrieval. *Information Processing and Management*. **34**(2/3): 191–218.

Egghe, L. and Rousseau, R. (1997). Duality in Information Retrieval and the Hypergeometric Distribution. *Journal of Documentation*. **53**(5), 488–496.

Egghe, L. and Rousseau, R. (1998). Topological Aspects of Information Retrieval. *Journal of the American Society for Information Science*. **49**(13): 1144–1160.

Ellis, D. (1966). The dilemma of measurement in information retrieval research. *Journal of the American Society for Information Science*. **47**: 23–36.

Everett, D.M. and Cater, S.C. (1992). Topology of document retrieval systems. *Journal of the American Society for Information Science*. **43**(10): 658–673.

Favela, J. and Meza, V. (1999). Image retrieval agent: integrating image content and text. *IEEE Intelligent Systems*, 14(5): http://computer.org/intelligent/ex1999

Feldman, J.-A. and Ballard, D.-H. (1988). Connectionist models and their properties. *In*: Anderson, J. A. and Rosenfeld, E. (eds.) *Neurocomputing Foundations of Research*. The MIT Press, Cambridge, Massachusetts, London, England, 481–508.

Florian, D. and Buckland, M.K. (1990). Information Retrieval: From Task Complexity to Artificial Intelligence. *In*: Ziinterhof, H. (ed.) *Proceedings of FIT'90 Conference*. Salzburg, Austria, Oldenbourg, Wien, 13–27.

Ford, N. (1991). Knowledge–Based Information Retrieval. *Journal of the American Society for Information Science*. **42**(1): 72–74.

Forsyth, F. and Rada, R. (1986). *Machine Learning: Applications in Expert Systems and Artificial Intelligence*, Ellis Harwood Ltd.

Frakes, W. B. and Baeza–Yates, R. (1992). *Information Retrieval: data structures and algorithms*. Prentice Hall, Englewood Cliffs, N.J.

Frakes, W. B. and Baeza–Yates, R. (1992). *Infromation Retrieval: Data Structures and Algorithms*, Prentice Hall, New Jersey.

Frants, V. and Shapiro, J. (1991). Algorithm for Automatic Construction of Query Formulations in Boolean Form. *Journal of the American Society for Information Science*, **42**(1): 16–26.

Frants, V.I., Shapiro, J. and Voikunskii, V.G. (1997). *Automated Information Retrieval: Theory and Methods*. Academic Press, San Diego.

Froehlich, T.J. (1991). Towards a better conceptual framework for understanding relevance for information science. In: *Proceedings of the American Society for Information Science*. Washington, DC, Medford, NJ, Learned Infomation, 118–125.

Froehlich, T.J. (1994). Relevance reconsidered – Towards an agenda for the 21$^{st}$ century. *Journal of the American Society for Information Science*. **45**(3): 124–133.

Fuhr, N. (1992). Probabilistic Models in Information Retrieval. *The Computer Journal*. **35**(3): 243–255.

Fuhr, N. (1999). Towards data abstraction in networked information retrieval systems. *Information Processing and Management*. **35**(2): 101–119.

Fuhr, N. and Rolleke, T. (1997). A probabilistic relational algebra for the integration of information retrieval and database systems. *ACM Transactions On Information Systems*. **15**(1): 32–66.

Fuhr, N. and Rolleke, T. (1998). HySpirit – a Probabilistic Inference Engine for Hypermedia Retrieval in Large Databases. *In*: Schek, H.J., Saltor, F., Ramos, I. and Alonso, G. (eds.) *Proceedings of the 6th International Conference on Extending Database Technology EDTB' 98*. Springer Verlag, 24–38.

Garey, M.R. and Johnson, D.S. (1979). *Computers and Intarctability: A guide to the theory of NP–completenss*. Freeman.

Ginsberg, A. (1993). A Unified Approach to Automatic Indexing and Information Retrieval. *IEEE Computer*. **8**(5): 46–56.

Goffmann, W. (1964). On relevance as a measure. *Information Storage and Retrieval*. **2**(3): 201–203.

Goffmann, W. (1970). A general theory of communication. *In*: Saracevic, T. (ed.) *Introduction to information science*. Bowker, New York, 726–747.

Goker, A. (1997). Context learning in Okapi. *Journal of Documentation*. 1: 80–83.

Goker, A. and McCluskey, T.L. (1991). Incremental learning in a probabilistic information retrieval system. *In*: Birnbaum, L. A. and Collins, G. C. (eds.) *Machine Learning, Proceedings of the 8$^{th}$ International Workshop ML91*. Morgan Kaufmann Publishers, Inc., 255–259.

Gordon, M. (1997). It's 10 A.M. Do you Know Where your Documents are? The Nature and Scope of Information Retrieval Problems in Business. *Information Processing and Management*. **33**(1): 107–121.

Gordon, M. and Pathak, P. (1999). Finding information on the World Wide Web: the retrieval effectiveness of search engines. *Information Processing and Management*. **35**(2): 141–180.

Gordon, M. D. and Lenk, P. (1991). A utility theoretic examination of the probability ranking principle in information retrieval. *Journal of the American Society for Information Science*. **42**: 703–714.

Gordon, M.D. (1990). Evaluating the Effectiveness of Information Retrieval Systems Using Simulated Queries. *Journal of the American Society for Information Science*. **41**(5): 313–323.

Gordon, M.D. and Kochen, M. (1989). Recall–Precision trade–off: A derivation. *Journal of the American Society for Information Science*. **40**: 145–151.

Gray, A. (1993). *Modern differential geometry of curves and surfaces*. CRC Press, Inc.

Guan, T. and Wong, K.–F. (1998). KPS: a Web information mining algorithm. http://www.decweb.ethz.ch/WWW8

Gudivada, V.N. and Raghavan, V.V. (1997). Modeling and Retrieving by Content. *Information Processing and Management*. **33**(4): 427–452.

Guinan, C. and Smeaton, A. F. (1992). Information Retrieval from Hypertext Using Dynamically Planned Guided Tours. *In*: Lucarella, D. et al. (eds.) *Proceedings of ECHT'92*. Milano, Italy, 122–130.

Halpern, M. I. and Shaw, C. J. (1969, eds.). *Annual Review in Automatic Programming 5*. Vol. 13, Pergamon Press.

Harary, F. (1972). *Graph Theory*. Addison–Wesley, Reading, Massachussetts.

Harel, D. (1998). Towards a Theory of Recursive Structures. *In*: Prim, L. et al.(eds.) *MFCS '98, LNCS 1450*. Springer Verlag Berlin Heidelberg, 36–53.

Harper, D.J. and Walker, A.D.M. (1992). ECLAIR, an extensible Class Library for Information Retrieval. *The Computer Journal*. **35**(3): 256–267.

Hays, D.G. (1966, ed.). *Readings in Automatic Language Processing*. Elsevier, New York.

Heine, M. H. (1999a). Reassessing and extending the Precision and Recall concepts. eWIC, British Computer Society, http://www.ewic.org.uk.

Heine, M. H. (1999b). Measuring the Effects of AND, AND NOT and OR Operators in Document Retrieval Systems Ulsing Directed Line Segments. In: Crestani, F. and Lalmas, M. (eds.) *Proceedings of the Workshop on Logical and Uncertainty Models for Information Systems*, University College, London, United Kingdom, 5th July, 55–76.

Heisenberg, W. (1971). *Physics and Beyond – Encounters and Concersations*. Harper and Row Publishers, Inc.

Himmelblau, D.M. (1972). *Applied Nonlinear Programming*. McGraw–Hill.

Huang, X.J. and Robertson, S.E. (1997). Application of probabilistic methods to Chinese text retrieval. *Journal of Documentation*. **53**: 74–79.

Huibers, T.W.C. and Bruza, P.D. (1996). Situations, a General Framework for Studying Information Retrieval. *In*: Leon, R. (ed.) *Information Retrieval New Systems and Current Research. Proceedings of the 16th Research Colloquim of the British Computer Society Information Retrieval Specialist Group*. Taylor Graham, Drymen, 3–25.

Huibers, T.W.C. and Denos, N. (1996). A Qualitative Ranking Method for Logical Information Retrieval Models. *Research Report*. LGI(CLIPS)–IMAG, Grenoble, France.

Hurt, C.D. (1998). Nonmonotonic Logic for use in Information Retrieval: An Exploratory Paper. *Information Processing and Management*. **34**(1): 35–41.

Hwang, Ch. L. and Masud, A. S. M. (1979). *Multiple Objective Decision Making – Methods and Applications*. Berlin, Heidelberg, New York.

Hwang, G. C. and Yoon, K. (1981). *Multiple Attribute Decision Making*. Berlin, Heidelberg, New York.

*IEEE Intelligent Systems* (1999). Vol. 14, No. 4, September/October.

Ingwersen, P. (1992). *Information retrieval interaction*. Taylor Graham, London.

Jaczynski, M. and Trousse, B (1998). WWW assisted browsing by reusing past navigations of group of users. *In: Proceedings of European Workshop on Case–based Reasoning*. 160–171.

Janes, J. W. (1994). Other people's judgements: A comparison of user's and others' judgements of docuement relevance, topicality and utility. *Journal of the American Society for Information Science*. 45(3): 160–171.

Jansen, B.J., Spink, A., Bateman, J. and Saracevic, T. (1998). Real Life Information Retrieval: A study of user queries on the Web. *SIGIR–Forum*. 32(1): 5–17.

JASIS (1994). Special topic issue: Relevance research. *Journal of the American Society for Information Science*. 45.

Kang, H.K. and Choi, K.S. (1997). Two–level Document Ranking Using Mutual Information In Natural Language. *Information Processing and Management*. 33(3): 289–306.

Kantor, P. et al. (1995). Combining the Evidence of Multiple Query Representations for Information Retrieval. *Information Processing and Management*. 31(3): 431–448.

Karamuftuoglu, M. (1998). Collaborative Information Retrieval: Toward a Social Informatics View of IR Interaction. *Journal of the American Society for Information Science*. 49(12): 1070–1080.

Kevin, L.F., Frieder, O., Knepper, M.M. and Snowberg, E.J. (1999). SENTINEL: A Multiple Engine Information Retrieval and Visualization System. *Journal of the American Society for Information Science*. 50(7): 616–625.

Kilgour, F.G., Moran, B.B. and Barden, J.R. (1999). Retrieval Effectiveness of Surname–Title–Word Searches for Known Items by Academic Library Users. *Journal of the American Society for Information Science*. 50(3): 265–270.

Kim, J. Y. and Shawe–Taylor, J. (1994). fast String Matching using an N–grams Algorithm. *Software – Practice and Experience*. 24(1): 79–88.

Kim, W.Y., Kim, H.H. and Lee, Y.J. (1998). probabilistic retrieval incorporating the relationship of descriptors incrementally. *Information Processing and Management*. 34(4): 417–430.

Kinnebrock, W. (1992). *Neuronale Netze*. Oldenbourg Wien.

Kneale, K. and Kneale, M. (1962). *The Development of Logic*. Oxford University Press.

Kochen, M. (1974). *Principles of information retrieval*. Melcille, Los Angeles, CA.

Kohle, M. (1990). *Neuronale Netze*. Springer Verlag.

Koll, M. and Srinivasan, P. (1990). Fuzzy versus probabilistic models for user relevance judgements. *Journal of the American Society for Information Science*. 41(4): 264–271.

Kolmogoroff, A. (1950). *Foundation of Probability*. New York.

Korfhage, R.R. (1997). *Information Storage and Retrieval*. JohnWiley and Sons, Wiley, New York.

Kowalski, G. (1997). *Information Retrieval Systems: Theory and implementation*. Kluwer Academic Publishers, Boston, MA.

Kraft, D. H. (1985). Advances in Information Retrieval: Where is That /#*%@^ Record? In: Yovits, M. (ed.), *Advances in Computers*, 24, New York, NY: Academic Press, 277–318.

Kraft, D. H. and Bookstein, A. (1978). Evaluation of information retrieval systems: a decision theory approach. *Journal of the American Society for Information Science*. 29: 31–40.

Kraft, D. H. and Buell, D. A. (1983). Fuzzy sets and generalized Boolean retrieval systems. *International Journal of Man–Machine Studies*, 19(1): 45–56.

Kraft, D. H., Bordogna, G. and Pasi, G. (1995). An extended fuzzy linguistic approach to generalizing Boolean information retrieval. *Journal of Information Sciences, Applications*, 2(3): 119–134.

Kraft, D. H., Petry, F. E., Buckles, B. P. and Sadasivan, T. (1994) Applying genetic algorithms to information retreival systems via relevance feedback. *In*: Bosc, P. and Kacprzyk, J. (eds.) *Fuzzy Sets and Possibility Theory in Datbase Management Systems, Studies in Fuzziness Series*. Physica Verlag, Heidelberg, Germany, 330–344.

Kraft, D.H. and Boyce, B.R. (1995). Approaches to Intelligent Information Retrieval. *In*: Petry, F.E. and Delcambre, M.L. (eds.) *Advances in Databases and Artificial Intelligence, vol. 1: Intelligent Database Technology: Approaches and Applications*. JAI Press, Greenwich, CT, 243–261.

Kraft, D.H. and Buell, D.A. (1992). Fuzzy Sets and Generalized Boolean Retrieval Systems. *In*: Dubois, D., Prade, H. and Yager, R. (eds.) *Readings in Fuzzy Sets for Intelligent Systems*. Morgan Kaufmann Publishers, San Mateo, CA.

Kraft, D.H. and Monk, D. (1998). Applications of Fuzzy Computation – Information Retrieval: A Case Study with the CASHE:PVS System. In: Ruspini, E., Bonissone, P. and Pedrycz, W. (eds.) *Handbook of Fuzzy Computation, Part G: Fuzzy Computation in Practice, G6: Information Science*. Oxford University Press and Institute of Physics Publishing, New York, NY.

Kraft, D.H., Bordogna, P. and Pasi, G. (1998). Fuzzy Set Techniques in Information Retrieval. *In*: Didier, D. and Prade, H. (eds.) *Handbook of Fuzzy Sets and Possibility Theory. Approximate Reasoning and Fuzzy Infomation Systems*. Kluwer Academic Publishers, AA Dordrecht, The Netherlands, Chp. 8.

Kwok, K.L. (1989). A Neural Network for the Probabilistic Information Retrieval. *In*: Belkin, N.J. and van Rijsbergen, C.J. (eds.) *Proceedings of the 12th Annual International ACMSIGIR Conference on Research and Development in Information Science*. ACM Press, Cambidge, MA, 21–29.

Kwok, K.L. (1995). A network approach to probabilistic information retrieval. *ACM Transactions On Information Systems*. 13(3): 243–353.

Lakatos, I. (1976). *Proofs and Refutations – The Logic of Mathematical Discovery*. Cambridge University Press.

Lalmas, M. (1998). Logical Models in Information Retrieval: Introduction and Overview. *Information Processing and Management*. 34(1): 19–33.

Lalmas, M. and van Rijsbergen, C.J. (1992). A logical model of information retrieval based on situation theory. *BCS 14th Information Retrieval Colloquium*. Lancaster, 1–13.

Lalmas, M. and van Rijsbergen, C.J. (1993). Situation theory and Dempster–Shafer's theory of evidence for information retrieval. *Proceedings of Workshop on Incompleteness and Uncertainty in Information Sytems*. Concordia University, Montreal, 62–67.

Lancaster, F. W. (1979). *Information retrieval systems: Characteristics, testing and evaluation* (2nd edition). John Wiley and Sons, New York.

Lee, J. (1995). Combining Multiple Evidence from Different Properties of Weighting Schemes. *In*: *Proceedings of the ACM SIGIR Conference on Research and Development in Information Retrieval*. Seattle, Washington, 180–188.

Lee, J.H. (1998). Combining the Evidence of Different relevance Feedback Methods for Information Retrieval. *Information Processing and Management*. 34(6): 681–691.

Lee, J.J. and Kantor, P.B. (1991). A Study of Probabilistic Information Retrieval Systems In the Case of Inconsistent Expert Judgement. *Journal of the American Society for Information Science.* **42**(3): 166–172.

Lin, X. (1997). Map Displays for Information Retrieval. *Journal of the American Society for Information Science.* **48**(1): 40–54.

Liu, G.Z. (1997). Semantic Vector Space Model: Implementation and Evaluation. *Journal of the American Society for Information Science.* **48**(5): 395–417.

Liu, Y. (1995). *Statistical and Knowledge Bases Approaches for Sense Disambiguities in Information Retrieval.* MSc, Department of Computing, University of Guelph, Guelph, Ontario.

Luhn, H.P. (1959). Keyword–in–Context Index for Technical Literature (KWIC Index). *In*: Hays, D. D. (ed.). *Readings in Automatic Language Processing.* American Elsevier Publishing Company, Inc., (1966), 159–167.

MacFarlane, A., Robertson, S.E. and McCann, J. (1997). Parallel computing in information retrieval – an updated review. *Journal of Documentation.* **53**: 274–315.

Maron, M E. (1977). On indexing, retrieval an the meaning of about. *Journal of the American Society for Information Science.* **28**: 38–43.

Maron, M.E. and Kuhns, J.–L. (1960). On relevance, probabilistic indexing and information retrieval. *Journal of the Association for Computing Machinery.* **7**: 219–244.

Martin, P. and Eklund, P. (1998). Embedding knowledge in Web documents. http://www.decweb.ethz.ch/WWW8

Martin, W.T. and Reissner, E. (1961). *Elementary Differential Equations.* Addison–Wesley, Reading, MA, London, England.

Meadow, C. T. (1988). Comment on Some Recent Comment on Information Retrieval. *SIGIR,* **22**(1–2):5–8.

Meadow, C. T. (1992). *Text Information Retrieval Systems.* Academic Press, New York.

Meadow, C.T., Boyce, B.R. and Kraft, D.H. (1999). *Text Information Retrieval Systems.* second edition, Academic Press, San Diego, CA.

Mechkour, M., Harper, D.J. and Muresan, G. (1998). The WebCluster Project: Using clustering for mediating access to the World Wide Web. *In: Proceedings of the ACM SIGIR International Conference on Research and Development in Information Retrieval.* Melbourne, Australia, 357–358.

Meghini, C., Sebastiani, F., Straccia, U. and Thanos, C. (1993). A model of information retrieval based on a terminological logic. *In: ACM SIGIR 16th International Conference on Research and Development in Information Retrieval.* ACM Press, Pittsburh, PA, New York, 298–307.

Mehtre, B.M., Kankahhali, M.S. and Lee, W.F. (1997). Shape Measures for Content Based Image Retrieval: A Comparison. *Information Processing and Management.* **33**(3): 319–336.

Mehtre, B.M., Kankahhali, M.S. and Lee, W.F. (1998). Content–Based Information Retrieval Using A Composite Color–Shape Approach. *Information Processing and Management.* **34**(1): 109–120.

Meinke, K. and Tucker, J. V. (1992). Universal Algebra. *In*: Abramsky, S. and Gabbay, D.M. and Maibaum, T.S.E. (eds.) *Handbook of Logic in Computer Science.* Vol. 1, Oxford Science Publications, Clarenden Press.

Melucci, M. (1998). Passaghe Retrieval: A probabilistic Technique. *Information Processing and Management.* **34**(1): 43–68.

Miettinen, K.M. (1999). *Nonlinear Multiobjective Optimization.* Kluwer Academic Publishers, Boston/London/Dordrecht.

Miller, G. (1990). Special Issue, WordNet: An on–line lexical database. *International Journal of Lexicography*, 3(4).

Miyamoto, S. (1990). *Fuzzy Sets in information retrieval and cluster analysis*. Kluwer Academic Publishers, Dordrecht.

Miyamoto, S. (1998). Application of Rough Sets to Information Retrieval. *Journal of the American Society of Information Science*. 49(13): 195–205.

Mizzaro, S. (1996). A cognitive analysis of information retrieval. In: Ingwersen, P. and Pors, N. O. (eds.), *Information science: Integration in Perspective – Proceedings of COLISZ*, Copenhagen, Denmark, The Royal School of Librarianship, 233–250.

Mizzaro, S. (1997). Relevance: The Whole History. *Journal of the American Society of Information Science*. 48(9): 810–832.

Mock, K.J. and Vemuri, V.R. (1997). Information Filtering via Hill Climbing, Wordnet and Index Patterns. *Information Processing and Management*. 33(5): 633–644.

Negoita, C. V. (1973). On the notion of relevance in information retrieval. *Kybernetes*. 2(3): 161–165.

Negoita, C. V. and Flondor, P. (1976). On fuzziness in information retrieval. *International Journal of man–Machine Studies*, 8(6): 711–716.

Nie, J. Y. (1988). An outline of a general model for information retrieval systems. In: *ACM SIGIR 11th International Conference on Research and Development in Information Retrieval*, Presses Universitaire de Grenoble, Grenoble, France, 495–506.

Nie, J. Y. (1989). An information retrieval model based on modal logic. *Information Processing and Management*. 25(5): 477–491.

Nie, J. Y. and Chiaramella, Y. (1990). A retrieval model based on an extended modal logic and its application to the rime experimental approach. In: *ACM SIGIR 13th International Conference on Research and Development in Information Retrieval*. Bruxelles, ACM Press, New York, 25–43.

Nie, J.Y. (1990). An information retrieval model based on modal logic. *Information Processing and Management*. 25(5): 477–491.

Nie, J.Y. (1992). Towards a probabilistic modal logic for semantic–based information retrieval. In: *ACM SIGIR 15th International Conference on Research and Development in Information Retrieval*, Copenhagen, Denmark, ACM Press, New York, 140–151.

Nie, J.Y., Brisebois, M. and Lepage, F. (1995). Information retrieval as counterfactual. *The Computer Journal*. 38(8): 643–657.

Oxley, J.G. (1992). *Matroid Theory*. Oxford University Press, Oxford New York Tokyo.

Pao, Y. H. (1989). *Adaptive Pattern Recognition and Neural Networks*, Addison–Wesley.

Paris, L.A.H. and Tibbo, H.R. (1998). Freestyle vs Boolean: A comparison of partial and exact match retreival systems. *Information Processing and Management*. 34(2/3): 175–190.

Park, H. (1997). Relevance of Science Information: Origins and Dimensions of Relevance and their Implications to Infromation Retrieval. *Information Processing and Management*. 33(3): 339–351.

Park, T. K. (1994). Toward a theory of user–based relevance: A call for a new paradigm of inquiry. *Journal of the American Society for Information Science*. 45: 135–141.

Pasi, G. (1999) A logical formulation of the Boolean model and weighted Boolean Model. In: Crestani, F. and Lalmas, M. (eds.) *Proceedings of Workshop on Logical and Uncertainty Models for Information Systems*. University College, London, United Kingdom, 5th July, 1–11.

Pasi, G. and Marques Pereira, R. A. (1999). A decision making approach to relevance feedback in information retrieval: a model based on a soft consensus dynamics. International Journal of Intelligent Systems. 14(1): 1–18.

Pazzani, M. and Billsus, D. (1997). Learning and revising user profiles: The identification of interesting Web sites. *Machine Learning*. 27: 313–331.

Pearce, C. and Nicolas, C. (1996). TELLTALE: Experiments in a Dynamic Hypertext Environment for Degraded and Multilingual Data. *Journal of the American Society for Information Science*. 47(4): 263–275.

Petry, F.E. and Kraft, D.H. (1997). Managing Uncertainty in databases and Information Retrieval Systems. *Fuzzy Sets and Systems*. 90: 183–191.

Petry, F.E., Buckles, B.P., Kraft, D.H., Prabhu, D. and Sadavisan, T. (1997). The use of genetic Programming to Build Queries for Information Retrieval. *In*: Baeck, T., Fogel, D. and Michalewicz, Z. (eds.) *Handbook of Evolutionary Computation*. Oxford University Press, New York.

Philips, I.C.C. (1992). Recursion Theory. *In*: Abramsky, S. and Gabbay, D.M. and Maibaum, T.S.E. (eds.) *Handbook of Logic in Computer Science*. Vol. 1, Oxford Science Publications, Clarenden Press.

Qui, Y. and Frei, H. P. (1993). Concept Base Query Expansion. *Proceedings of the Sixteenth Annual International ACM SIGIR Conference on Research and Development in Information Retrieval*, 160–169.

Radecki, T. (1976). Mathematical model of information retrieval system based on the concept of fuzzy thesaurus. *Information Processing and Management*. 12: 131–318.

Radecki, T. (1977). Mathematical model of time–effective information retrieval system based on the theory of fuzzy sets. *Information Processing and Management*. 13: 109–116.

Radecki, T. (1979). Fuzzy set theoretical approach to document retrieval. *Information Processing and Management*, 15(5): 247–260.

Radecki, T. and Kraft, D. H. (1985). Incorporating relevance feedback into fuzzy information retrieval procedures. In: *Proceedings of the First IFSA International Fuzzy Systems Association Congress*. Palma de Mallorca, Spain.

Ragade, R. K. and Zunde, P. (1974). Incertitude characterization of the retriever–system communication process. In: *Proceedings of the ASIS 37th Annual Meeting*, Atlanta, GA, Washington, DC, 13–17 October, 11, 128–129.

Raghavan, V.V. and H. Sever (1995). On the reuse of past optimal queries. *In*: *ACM SIGIR International Conference on Research and Development in Information Retrieval*. 344–350.

Raghavan, V.V. and Wong, S.K.M. (1986). A Critical Analysis of Vector Space Model for Information Retrieval. *Journal of the American Society for Information Science*. 37: 279–287.

Raghavan, V.V., Wong, S.K.M. and Ziarko, W. (1990). Vector Space Model in Information Retrieval. *Encyclopedia of Computer Science and Technology*. 22: 423–446.

Recski, A. (1989). *Matroid Theory and Its Applications*. Akademiai Kiadó, Budapest.

Reddaway, S. F. (1991). High speed text retrieval from large databases on a massively parallel processor. *Information Processing and Management*. 27(4): 311–316.

Robertson, S.E. and Spark Jones, K. (1976). Relevance weighting of search terms. *Journal of the American Society of Information Science*. 27: 129–146.

Robertson, S.E., Maron, M.–E. and Cooper, W.S. (1982). Probability of relevance: A unification of two competing models for document retrieval. *Information Technology: research and Development*. 1: 1–21.

Rocchio, J. J. (1971). Relevance Feedback in Information Retrieval. *In:* Salton, G. (ed.) *The SMART Storage and Retrieval System.* Prentice Hall, Englewood Cliffs, NJ, 313–323.

Roland, O. (1994). *The interpretation of quantum mechanics.* Princeton University Press, Princeton, New Jersey.

Rossopoulos, N., Kelley, S. and Vincent, F. (1995). Nearest neighbor queries. *In: Proceedings of the 1995 ACM SIGNOID Interantional Conference on Management of Data.* San Jose, CA, 71–79.

Rousseau, R. (1998). Jaccard Similarity Leads to the Marczewski–Steinhaus Topology for Information Retrieval. *Information Processing and Management.* **34**(1): 87–94.

Salton, G. (1965). Automatic Phrase Matching. *In:* Hays, D.D. (ed.). *Readings in Automatic Language Processing.* American Elsevier Publishing Company, Inc., (1966), 169–189.

Salton, G. (1968). *Automatic Information Organisation and Retrieval.* McGraw Hill, New York.

Salton, G. (1971a) Relevance Feedback and the optimization of retrieval effectiveness. *In:* Salton, G. (ed.) *The SMART Storage and Retrieval System.* Prentice Hall, Englewood Cliffs, NJ, 324–336.

Salton, G. (1971a). *The SMART Retrieval System – Experiment in Automatic Document Processing.* Prentice–Hall, Englewood Cliffs, New Jersey.

Salton, G. (1989). *Automatic text processing.* Addison Wesley, Reading, MA.

Salton, G. and Buckley, C. (1990). Improving retrieval performance by relevenace feedback. *Journal of the American Society of Information Science.* **41**: 288–297.

Salton, G., Allan, J. and Singhall, A. (1996). Automatic Text Decomposition and Structuring. *Information Processing and Management.* **32**(2): 127–138.

Salton, G., Fox, E. A. and Voorhees, E. (1985). Advanced feedback methods in information rerieval. *Journal of the American Society of Information Science.* **36**: 200–210.

Salton, G., McGill, M. (1983). *Introduction to Modern Information Retrieval.* McGraw Hill, New York.

Salton, G., Singhall, A., Mitra, M. and Buckley, C. (1997). Automatic Text Structuring and Summarization. *Information Processing and Management.* **33**(2): 193–207.

Saracevic, T. (1975). Relevance:A review of and a framework for the thinking of the notion in information retrieval. *Journal of the American Society of Information Science.* **26**: 321–343.

Saracevic, T. and Kantor, P. (1988a). A study of information seeking and retrieving. II. Users, questions and effectiveness. *Journal of the American Society of Information Science.* **39**: 177–196.

Saracevic, T. and Kantor, P. (1988b). A study of information seeking and retrieving. III. Searchers, searches and overlap. *Journal of the American Society of Information Science.* **39**: 197–216.

Savoy, J. (1993). Searching information in hypertext systems using multiple sources of evidence. *International Journal of Man–Machine Studies.* **38**: 1017–1030.

Savoy, J. (1995). A New Probabilistic Scheme for Information Retrieval in Hypertext. *The New Review of Hypermedia and Multimedia.* **1**: 107–115.

Savoy, J. (1997). Statistical Inference in Retrieval Effectiveness Evaluation. *Information Processing and Management.* **33**(4): 495–512.

Schafer, G. (1976). *A mathematiocal theory of evidence.* Princeton University Press, Princeton, NJ.

Schafer, G. (1987). Belief functions and possibility measures. *In:* Bezdek, J. (ed.) *The Analysis of Fuzzy Information* (vol. 2). CRC Press, Boca Raton, FL.

Schamber, L. (1994). Relevance and information behavior. *Annual Review of Information Science and Technology.* **29**: 3–48.

Schutze, H. and Pedersen, J.O. (1997). A Cooccurrence–based Thesaurus and two Applications to Information Retrieval. *Information Processing and Management.* **33**(3): 307–317.

Sebastiani, F. (1994). A probabilistic termonological logic for information retrieval. *In*: *ACM SIGIR 17th International Coneference on Research and Development in Information Retrieval.* Springer Verlag, Dublin, Ireland, London, 122–130.

Sebastiani, F. (1998). Trends in...A critical review: On the Logic in Information Retrieval. *Information Processing and Management.* **34**(1): 1–18.

Seidl, T. and Kriegel, H. P. (1997). Efficient user–adaptable similarity search in large multimedia databases. *In*: *Proceedings of the 23rd VLDB International Conference.* Athens, Greece, 506–515.

Shannon, C. E. and Weaver, W. (1949). *The Mathematical Theory of Communication.* University of Illinois Press.

Shaw, W.M., Burgiu, R. and Howell, P. (1997). Performance Standards and Evaluation In Information Retrieval Test Collection: Cluster–Based Retrieval Models. *Information Processing and Management.* **33**(1): 1–14.

Shaw, W.M., Burgiu, R. and Howell, P. (1997). Performance Standards and Evaluation In Information Retrieval Test Collection: Vector–Space and Other Retrieval Models. *Information Processing and Management.* **33**(1): 15–36.

Smeaton, A. F. (1984). Relevance feedback and a fuzzy set of search terms in an information retrieval system. *Information Technology: Research Development Applications.* **3**(1): 15–23.

Smeaton, A. F. (1996). An Overview of Information Retrieval. *In*: Agosti, M. and Smeaton, A. F. (eds.) *Information Retrieval and Hypertext.* Kluwer Academic Publishers.

Smeaton, A. F. and van Rijsbergen, C. J. (1983). The Retrieval Effects of Query Expansion on a Feedback Document Retrieval System, *The Computer Journal*, 26(3): 239–246.

Smeaton, A.F. and Crimmins, F. (1999). Using A Data Fusion Agent for Searching the WWW. *http://decweb.ethz.ch/WWW6/Posters/752/FUSION–W.HTM*

Smyth, M.B. (1992). Topology. *In*: Abramsky, S. and Gabbay, D.M. and Maibaum, T.S.E. (eds.) *Handbook of Logic in Computer Science.* Vol. 1, Oxford Science Publications, Clarenden Press.

Soergel, D. (1994). Indexing and retrieval performance: The logical evidence. *Journal of the American Society of Information Science.* **45**: 589–599.

Spink, A. and Saracevic, T. (1997). Interaction in Information Retrieval: selection and Effectiveness of Search Terms. *Journal of the American Society of Information Science.* **48**(9), 741–761.

Spink, A., Goodrum, A. and Robins, D. (1998). Elicitation Behavior During mediated Information Retrieval. *Information Processing and Management.* **34**(2/3): 257–273.

Spink, A., Greisdorf, H. and Bateman, J. (1998). From Highly Relevant to Not Relevant: Examining Different Regions of Relevance. *Information Processing and Management.* **34**(5): 599–621.

Strzalkowski, T. (1995). Natural Language Information Retrieval. *Information Processing and Management.* **31**(3): 397–417.

Su, L. T. (1992). Evaluation measures for interactive infromation retrieval. *Information Processing and Management.* **28**(4): 503–516.

Su. L.T. (1994). The relevance of recall and precision in user evaluation. *Journal of the American Society of Information Science.* **45**(3), 207–217.

Su. L.T. (1998). Value of Search Results as a Whole as the Best Single Measure of information Retrieval Performance. *Information Processing and Management.* **34**(5): 557–579.

Sun, Q,. Shaw, D. and Davis, C.H. (1999). A Model for Estimating the Occurrences of same–Frequency Words and the Boundary between High– and Low–Frequncy Words in Texts. *Journal of the American Society for Information Science.* **50**(3): 280–292.

Tang, R. and Solomon, P. (1998). Toward an Understanding of the Dynamics of Relevance Judgement: An analysis of one person's search behavior. *Information Processing and Management.* **34**(2/3): 237–256.

Tarski, A. (1956). Fundamental concepts of the methodology of the deductive sciences. *In: Logic, Semantics, Metamathematics. Papers from 1923 to 1938.* Clarendon Press, Oxford, 60–109.

Tiamiyu, A. M. and Ajiferuke, I. Y. (1988). A total relevance and document interaction effects model for the evaluation of information retrieval processes. *Information, Processing and Management.* **24**(4): 391–404.

Tudhope, D. and Taylor, C. (1997). Navigation via Similarity: Automatic Linking Based on Semantic Closeness. *Information Processing and Management.* **33**(2): 233–242.

Turtle, H. R. and Croft, W. B. (1992). A comparison of text retrieval models. *The Computer Journal,* **35**(3): 279–290.

Ullman, J. D. (1980). *Principles of Database Systems.* Computer Science Press, Inc.

Valery, I.F., Shapiro, J. and Voiskunski, V.G. (1997). *Automated Information Retrieval: Theory and Methods.* Academic Press, San Diego, CA.

van Rijsbergen, C. J. and Lalmas, M. (1996). Infromation calculus for information retrieval. *Journal of the American Society for Information Science.* **47**: 385–398.

van Rijsbergen, C. J. (1996). Quantum Logic and Information Retrieval. *In:* Crestani, F. and Lalmas, M. (eds.) *Proceedings of Workshop on Logical and Uncertainty Models for Information Retrieval.* University of Glasgow, Glasgow, Scotland, July, pp. 1–2.

van Rijsbergen, C.J. (1977). A theoretical basis for the use of cooccurrence data in information retrieval. *Journal of Documentation.* **33**: 106–119.

van Rijsbergen, C.J. (1979). *Information Retrieval.* Butterworth, London.

van Rijsbergen, C.J. (1986a). A New Theoretical Framework for Information Retrieval. *SJGIR Forum,* **21**(1–2): 23–29.

van Rijsbergen, C.J. (1986b). A non–classical logic for information retrieval. *The Computer Journal.* **29**(6): 481–485.

van Rijsbergen, C.J. (1989). Towards an Information Logic. *In:* Belkin, N.J. and van Rijsbergen, C.J. (eds.) *Proceedings of the Twelfth Annual International ACMSIGIR Conference on Information Retrieval.* Cambridge, Massachusetts, USA, June 25–28, 77–86.

van Rijsbergen, C.J. (1992). Probabilistic Retrieval Revisited. *The Computer Journal,* **35**(3): 291–298.

van Rijsbergen, C.J. and Lalmas, M. (1996). An Information Calculus for Information Retrieval. *Journal of the American Society for Information Science.* **47**(5): 385–398.

von Neumann, J. (1927). Mathematische Grundlagen der Quantumphysik. *Göttinger Nachrichten, Math.–Phys.* Klasse, 1–46

Voorhees, E. M. and Gupta, N. K. and Johnson–Laird, B. (1995). Learning collection fusion strategies. *In: Proceedings of the 18th ACM SIGIR Conference on Research and Development in Information Retrieval.* Seattle, 172–179.

Voorhees, E. M., Gupta, N. K. and Johnson–Laird, B. (1994). The collection fusion problem. *In: Proceedings of the Third TREC–3 Conference.* 95–104.

Vorhees, E. (1994). Query Expansion Using Lexical–Semantic Relations. *Proceedings of the Seventeenth Annual International ACM SIGIR Conference on Research and Development in Information Retrieval*, 61–69.

Waller, W. G., and Kraft, D. H. (1979). A mathematical model of a weighted Boolean retrieval system. *Information Processing and Management*, 15: 235–245.

Warren R.G. (1998). A Theory of Term Weighting Based on Exploratory data Analysis. *In: Proceedings of ACM SIGIR*. 11–19.

Watanabe, Y., Okada, Y., Kaneji, K. and Sakamoto, Y. (1999). Retrieving related TV news reports and newspaper articles. *IEEE Intelligent Systems*, 14(5): http://computer.org/intelligent/ex1999

Watters, C. (1999). Information Retrieval and the Virtual Document. To appear: *Journal of the American Society for Information Science*.

Watters, C.R. (1989). Logic Framework for Information Retrieval. *Journal of the American Society for Information Science*. 40(5): 311–324.

Weiss, M.A. (1995). *Data Structures and Algorithm Analysis*. The Benjamin/Cummnings Publishing Company, Inc.

Willard, S. (1970). *General Topology*. Addison–Wesley, Reading, MA.

Wolfram, D. and Dimitroff, A. (1998). Hypertext vs Boolean–Based Searching in a Bibliographic Database Environment: A Direct Comparison of Searcher Peformance. *Information Processing and Management*. 34(6): 669–679.

Wong, S.K.M. and Yao, Y.Y. (1990). A Generalized Binary Probabilistic Independence Model. *Journal of the American Society for Information Science*. 41: 324–329.

Wong, S.K.M. and Yao, Y.Y. (1991). A Probabilistic Inference Model for Information Retrieval based on Axiomatic Decision Theory. *Information Systems*. 16: 301–321.

Wong, S.K.M. and Yao, Y.Y. (1993). A Probabilistic Method for Computing Term–by–Term Relationship. *Journal of the American Society for Information Science*. 44: 431–439.

Wong, S.K.M. and Yao, Y.Y. (1995). On modeling information retrieval with probabilistic inference. *ACM Transactions On Information Systems*. 13(1): 38–68

Wong, S.K.M., Bollmann, P. and Yao, Y.Y. (1991). Information Retrieval based on Axiomatic Decision Theory. *General Systems*. 19: 101–117.

Wong, S.K.M., Butz, C.J. and Xiang, Y. (1995). A Method for Implementing a Probabilistic Model as a Relational Database. *In: Proceedings of the 11th Conference on Uncertainty in Artificial Intelligence*. Montreal, 156–164.

Wu, J.K. and Narasimhalu, D. (1998). Fuzzy Content–Based Retrieval in Image Databases. *Information Processing and Management*. 34(5): 513–534.

Yager, R. R. (1987). A note on weighted queries in information retrieval systems. *Journal of the American Society for Information Science*, 38(1): 23–24.

Yager, R.R. and Rybalov, A. (1998). On the Fusion of Documents from Multiple Collection Information Retrieval Systems. *Journal of the American Society for Information Science*. 49(13): 1177–1184.

Yang, J. J. and Korfhage, R. R. (1992). Query modification using genetic algorithms in vector space models. *Research Report* LIS045/IS92001, School of Library and Information Science, University of Piitsburgh, Pittsburgh, PA.

Yao, Y. Y. and Wong, S. K. M. (1991). A probabilistic inference model for Information Retrieval. *Information Systems*. 16(3): 301–321.

Yu, C. T., Meng, W., and Park, S. (1989). A Framework for Effective Retrieval. *ACM Transactions son Database Systems*, 14(2): 147–167.

Yu, C.T. and Salton, G. (1976). Precision weighting – an effective automatic indexing method. *Journal of the Association for Computing Machinery*. 23: 76–88.

Yu, C.T., Meng, W. and Park, S. (1989). A Framework for Effective Retrieval. *ACM Transactions on Database Systems* **14**: 147–167.

Zadeh, L. A. (1984). Fuzzy sets and applications. *In*: Yager. O et al. (eds.) John Wiley and Sons, Inc., 29–44.

Zadeh, L. A. (1989). Knowledge representation in Fuzzy Logic. *IEEE Transactions on Knowledge and Data Engineering.* **1**: 89–100.

Zadeh, L.A. (1984). Fuzzy sets and Applications. *In*: Yager, O. et al. (eds.) *Selected Papers by L.A Zadeh.* John Wiley and Sons, Inc., 29–44.

Zadeh. L.A. (1989). Knowledge representation in Fuzzy Logic. *IEEE Transactions on Knowledge and Data Engineering.* **1**: 89–100.

Zimmerman, H.J. (1996). *Fuzzy Set Theory – and Its Applications.* 3rd edition. Kluwer Academic Publishers, Boston/Dordrecht/London.

Zipf, G. (1949). *Human behavior and the principle of least effort.* Addison–Wesley, Cambridge, MA.

# Index